Outdoors With Greg Clark

Also by Gregory Clarkin Totem Books
THE BIRD OF PROMISE

Outdoors With Gregory Clark

A TOTEM BOOK
TORONTO

First published, 1971 by
McClelland and Stewart Limited, Toronto

This edition published 1976
by TOTEM BOOKS
a division of
Collins Publishers
100 Lesmill Road, Don Mills, Ontario

Syndicated by
Canada Wide Feature Service Ltd., Montreal
between 1948 and 1967 inclusive

Canadian Cataloguing in Publication Data

Clark, Gregory, 1892-1977.
Outdoors with Greg Clark

A selection of stories that appeared between
1948 and 1962 in the author's syndicated column,
"Gregory Clark's Packsack."

ISBN 0-00-216680-1 pa.

1. Natural history — Outdoor books. I. Title.

QH81.C52 1977 500.9 C77-001551-4

Photographs courtesy of
Louis Jacques

Printed in Canada by
Universal Printers Ltd., Winnipeg, Manitoba.

Contents

SUMMER

AUTUMN

WINTER

Foreword

The short pieces in this book are taken from several thousand written for a newspaper feature which was published five times a week for nearly fourteen years under the heading "Gregory Clark's Packsack." The Packsack dealt with every subject that caught Greg's fancy, but once a fortnight, and sometimes more frequently, he wrote about wild nature and shared his great store of knowledge and his feelings about nature in a manner that has rarely been duplicated elsewhere in his writing during the past twenty-five years.

When he wrote these pieces Greg Clark was an internationally recognized naturalist and conservationist. At the same time he was a sportsman. He was, indeed, a dean of sportsmen, renowned as a hunter, beagler and bird gunner in the regions he frequented in Ontario, and as an uncannily expert fly fisherman wherever he went.

As a sportsman, he examines the deep atavistic drives that send city men forth in the fall with rifles. As a sportsman, too, he ponders the problem of sharing the forest and wild life with thousands of non-hunters who don't want to kill but to enjoy the company of living wild creatures. As a sportsman/naturalist he is the advocate of all hungry nature, man included. His plea against man is that man wants too much.

The company Greg Clark keeps, as all his readers know, is lively and invariably entertaining. The company here, mostly birds and animals, is no exception.

HUGH SHAW

Introduction

It was the great humourist Will Rogers who forty years ago remarked: "We are all ignorant, only on different subjects."

That fascinated me. To what degree are we all in our right minds? Are we all loony in various degrees? The hundred percenters are confined to institutions. But we who are only five or ten percenters may be stamp collectors or dry-fly fishermen or protest marchers of one kind and another. How about colour blindness? There are means of certifying that a person is colour blind. But I know of no way of measuring the relative degree of sensibility of ten persons set in a room facing a bowl of beautiful multi-coloured flowers. I have always doubted that we all see the same thing in almost any situation. How about the connoisseurs of modern art?

My father was tone deaf. He also loved singing, loudly. In church as a small boy I had to stand beside him when the choir and congregation rose to sing. In the pew next behind ours there always sat two elderly spinsters dressed in funereal black. They too were tone deaf. And when the grand organ blasted out the opening chords of "Rock of Ages," you can't imagine the uproar of these three tone-deaf but loud and worthy Christians. There is no slide rule by which to measure the comparative sensibility of those who are exalted by the music of Mozart, Beethoven, Brahms, and those who are sent into the jerks by hard-rock bands.

Now what has all this to do with this small book in your hands? I was born with an insatiable curiosity about the natural world. It may be my form of lunacy. I have hunted, fished and prowled the woods and meadows since childhood. I have killed. And I have also stood transfixed in adoration. I soon found out that ladies who tended the most beautiful gardens didn't know one tree from another. And that the world was full of people who, when a splendid bird flew across their vision, were as little interested as if an autumn leaf had blown by.

In the long years that I pursued my trade as a newspaperman, I never failed an opportunity to sneak in a little paragraph about my passion for the out-of-doors. Not a grand passion, I agree; but comfortable. I was waving signals to my fellow lovers of the wild, and hoping to catch the eye of new fellow loonies and the blind and the deaf. This was long before the word ecology appeared in the media. Some of my editors even thought the word conservation had political overtones, and blue-pencilled it.

Then along came Hugh Shaw, my editor for many years; and in one of our conversations he mentioned the enormous rise in the public interest in conservation, the ecology, and the natural world in every aspect.

"Why don't you," he suggested "gather a little book of your writing about nature, outdoor sport and so on?"

"Why don't you?" I retorted.

And by golly, he got to work and sifted through hundreds of thousands of words to find these small pieces, which are not literature, but what the literary people call *obiter dicta*. That means things mentioned in passing.

So, come on in.

No!

Come on OUT.

GREGORY CLARK

Outdoors With Greg Clark

SPRING

Season of Peril

Across the immensity of Canada at the end of March a hundred thousand rivers, streams and brooks are poised ready on the brink of their flood.

Any day they will let go. And for a savage week or so, they will hurl the countless tons of ragged ice, the pitiless surge of their leaden water, against all that stands before them. No man-made force, not dynamite, not bulldozers, can hold a candle to the might of rivers.

Yet, when the raging tumult of winter's surrender is over, out of that blasted and turbulent water fragile mayflies will rise in the quiet sunset. Little trout, with the wide-eyed look of creatures to whom nothing has happened, will dart in the clearing stream. Out of the mud, strewn and flung along the banks, green shoots will almost instantly appear, and flowers as frail and lovely as dreams will in no time be standing there, unperturbed.

At this same time, while rivers go mad, millions of creatures are making ready to be born: muskrats, bears, mink, mice, moles, voles, deer, foxes, wolves, beaver, moose. Some of these are born amidst the thunder of the flood. Most of them will see daylight in April, but as we think of this annual fury of nature, we have to remember all the creatures already alive, and awaiting delivery into so strange a world. Their mothers have to stay very still and cautious while winter stages its shouting and battling death scene.

This is the time of greatest peril to wild nature in the whole year.

Frog Chorus

In any part of Canada where winter has relaxed its hold to the extent of even a modest puddle in a field or a creek flooding into the woods, frogs are welcoming spring with music that thrills the heart of country dwellers.

As fast as the ice lets go of the earth, the tiny spring peepers, hundreds of them to an acre of flooded roadside or swamp, come out of their winter hiding and raise their incredible voices in song. They keep it up for several weeks, and their music is one of the most warming of all spring songs to truck drivers, farmers and others who travel the roads at night.

The spring peeper's stage-name is Hyla crucifer, the cross-bearer. It is so fragile a creature that it appears semi-transparent. On its back is a dark X. When it blows up its throat to utter its far-reaching and melodious song, the bulge appears to be a huge pearl.

Spring peepers are not difficult to locate with a flashlight. Their music ceases instantly as you draw near their pond, but if you keep perfectly still, it begins tentatively again, and soon is in full choir. Spot your flashlight not on the water but on the grass stems and twigs. And there you will see the pearl of the cross-bearer as he sings his hymn to the spring.

In most parts of Canada the common toad is next in line among spring's welcomers. A great many people are wholly unaware that the soft, sweet musical thrill they can hear among the other night sounds of early spring is the love-song of the warty toad. Spring is well advanced before the other frogs, such as the leopard and the green, those familiar frogs of the summer beaches, get their music into the act.

Sometimes the spring night is fairly ringing with the clamour of spring peepers and toads. The air seems as filled with sounds as the sky is filled with stars. And when you stop to consider that every one of these hundreds or thousands of creatures within your hearing, within a radius of a few hundred yards of you, has survived the iron winter with all its zero temperatures and its tempests and ice, a fresh impression of the vitality of wild nature reaches your mind. These countless tiny frogs and clumsy toads, adults all, have wintered happily, painlessly, under stones, under logs, in frozen mud, under ice, all serene, while outside their ken the whole world shuddered.

Flustered Messenger

All of a fluster, like a messenger full of exciting news, the season's first robin comes running. Down he slopes on flirting wings and runs ten paces before he stops. He stands, looking like a courier whom neither snow nor rain, nor heat, nor gloom of night could stay. With head raised, as though glancing about for the audience to hear his news, he pauses. About him is an air of weariness and a little disarray, such as the herald from Marathon might have worn; a little dazed, as though speechless with the import of his message.

Other birds we see for the first time are usually in trees. They are en route elsewhere. They are passers-by whom we can greet with such pleasure as the dawn of spring arouses in us.

But the robin conveys a personal message. He has come home. And the word he brings is exciting to him and us both.

Despite the steady increase in human population in Canada, the number of people who will greet, much less notice the first robin singing, appears to grow smaller every year.

There was a time less than two generations ago when almost everybody in city and town waited for the first robin in a sort of general awareness. When you heard it, you called your next-door neighbours out to hear it too. It was important. As you walked home at supper time, you paused to listen to it. You smiled at strangers and pointed to a rooftop. People came and stood at their doors or put their heads out windows to hark.

That, of course, was back in the days when there was not so much noise in the world and not so many other interests and distractions. More people walked. Motor cars were few, and had not begun to fill the air with a substratum or mattress of ceaseless basic sound both day and night. The world was a dull, poorly lighted, slow-going institution in which a robin's first song had a chance to stand out, clear and distinct. Even a world-shaking cataclysm came silently off the newspaper page to you. Now it is shouted at you, even above the din of your car, as you race home and into your drive, to slam the car door and bound within for a quick supper. . . .

The robins sing as well as ever. But they are not wired to compete with the daily urban uproar.

Are there two races of robins? Is there one race that is domestic, and comes fearlessly right into the midst of our cities and towns, building its nests on the ledges of factories, on the porches of our homes and in trees along the edges of schoolyards loud with the tumult of children?

And another race that dwells in the wilderness, a shy, furtive, elusive bird that appears to be terrified of the human form? It flits far and fast amid the treetops uttering its familiar alarm note. Every town dweller knows this low, brief alarm call. Before the snow has gone, it makes magic in our gardens. But it is a curious experience to be walking in the wilds and hear that same call uttered with a true ring of alarm and to see the robins, usually in companies of half a dozen, flying as wild as doves and heading for places unknown in obvious fear of their lives.

By ringing birds with identification tags, ornithologists have proved that many birds return unerringly to their previous territories. It may well be that robins which return to their former nesting areas in city and town are the descendants of those that occupied the territory before the city or town was there, and thus have become adjusted to the disturbances and distractions of civilization through their rapid annual generations; and that the wilder robins encountered in the woods are the descendants of the normal or natural robins that have returned, century after century, to their ancestral areas in the forest.

As a trout fisherman, who sees and hears the forest robins in spring, I can assure you that they sing no more sweetly than the city robins. Indeed, it has sometimes struck me that the city robins sing more sweetly; but that is probably sheer sentimentality.

Spring Ice

When the ice goes out of a lake or pond, there occurs a violence which is far more important than the mere churning of broken ice by wind and wave. Usually, a little before the break-up, water accumulates on top of the ice. Some of it is melted ice, some may be rain or snow. As much as six inches to a foot of water can form on top of the ice. And the day comes when this burden grows a little too heavy for the strength of the ice-lid on the lake. Around the shores, the ice weakens. And that vast weight of water begins to rush down under the ice, while the ice itself struggles to rise and stay on top.

This powerful current of water, rushing to escape down under the ice all around the margins of the lake, carries all manner of sand, silt, vegetation with it, away from the shores, creating sand beaches, bars and shoals far out beyond the normal beat of waves in summer. Many of the best beaches on fresh water were built by this sucking under of water off the ice before the break-up.

When the break-up comes, the water on the surface is the coldest water in the lake and is trying immediately to sink down to the bottom, where it belongs. If there is a breeze to help, the cold surface water, filled with oxygen, is rolled to the shores where it instantly folds under the warmer water and slides away to the deepest depth it can find, according to thermal laws. This turnover of the water of lake and pond is absolutely vital to all life in it. Oxygen is distributed to far greater depths than wind and wave could ever penetrate.

When we go up to our summer homes and find the wharf all heaved and askew, it was not merely the grinding of chunks of ice that did the trick. It was massive weights of water and massive weights of ice heaving and lifting in normal competition. And the limpid water we look down into off our familiar docks is water rolled up from dark and lifeless depths by the sliding under of the icy surface water during the spring break-up.

Give and Take

We all have to feed one another. We feed the cows and chickens, and they feed us. In nature, it is all give and take. Of course, the vegetarians feel a little superior to the rest of us – which is possibly why they are vegetarians. We have all got to feel superior to somebody, even if only in humility. But the vegetarians would soon starve if the farmers did not feed the soil. And you know what they feed the soil with. . . . Maybe there are vegetarians who restrict their food strictly to uncultivated vegetable matter, such as grows wild. These would be the real Covenanters.

But the balance of nature is like the balance of the sea, and all the water of the earth that runs at last down to it. A rabbit eats nothing but vegetable matter and therefore must be looked upon as one that preys on nobody. But if there were no foxes or owls to eat the rabbits, the rabbits would in due time multiply in such immense numbers that they would eat down all the vegetation and perish in a cataclysmic famine. In which case, the rabbits would, to all intents and purposes, and by a roundabout means, have eaten each other. Cannibals. It is the fox and the owl and many another animal besides man who preserve the rabbit from that monstrous fate.

The young rabbits are already born in April. So are the young owls. They are the early comers. The young rabbits will be hopping about just in time for the young owls to find as they venture from their nests. By the time the young foxes are out foraging nervously for their first supper away from home, the vegetation will be up and the woods dense. But from the nests, by then, the young thrushes will be falling, and the young grosbeaks and warblers. Nature arranges everything very comfortably for all of us and does her best to keep us in our place. It is only man who has got a little out of hand.

Survival of the Stupid

Early in the spring, before the leaves had formed their first mist of green over the woods, I was in a canoe on a small river and heard a partridge drumming not a hundred yards to my left. My eye happened to catch sight of a hawk soaring less than 200 feet overhead. It was not an accipiter, such as are supposed to prey on game birds, but either a red-tailed or a broad-wing hawk, the buteos that deal mostly in mice, snakes and insects.

With a few fast flaps, the buteo turned and then set its wings half closed, its shoulders braced, and, assuming the heroic hawk posture for the stoop, dove steeply towards the still drumming partridge. A few minutes later, I found the partridge feathers that indicated one little wild rooster had strutted his last love call.

Later, when the forest had acquired its veil of green, I encountered a fox carrying a partridge. It too was a game cock, doubtless interrupted in the valiant act of drumming. The foliage which, as the season advances, screens the drummers from the hawks, rises out of the ground as well as forming overhead, and so serves as cover for the stealthy approach of the fox and other predators that will spend two or three hours stalking a drumming bird, advancing while he drums, hiding while he rests.

It is tough being a partridge and in love.

The wonder is that the race of partridge was not extinguished ages ago. Nature seems to have been very ill-advised, making so many of her creatures so stupid and vulnerable. In the vainglory of the breeding season, so many species – the deer, the moose, the pheasant, the partridge – give themselves away to their enemies.

Yet the stupid survive and multiply. It is true amongst us humans, too. Maybe stupidity is the ideal towards which nature strives.

Winds of March

What do you suppose the winds of March are for? At no other season of the year do the trees appear to wrench and twist and sway as they do in the March gales. And that, of course, is what the winds of March are for; to carry from those topmost branches, moving, in a really big tree, twenty and thirty feet in their gyrations, down through the main branches and into the trunk itself, the torque, the pull and vibration of the gales, into the earth itself. For the motion you see in the topmost branches is being conveyed far deeper than the trunk, down into the roots, which loosen and till the soil upon which the tree feeds. Winter's grip, rigid at the frost line, is being loosened. The soil for yards around is being shaken, the moisture is being allowed to penetrate to depths, the sodden earth is being bunted and jerked so that the tiny thread-like rootlets, which are the true feeding mechanism of the tree, can move and reach out, in their miraculous way, into newly broken ground.

The large trees are doing a job of cultivation, with the wind's help, for a far larger economy than their own. Let us not for a minute suppose that the wind lends itself only to the trees. If we take our eyes off the spectacular March calisthenics of the trees, we can see that every little bush is whipping and lashing as furiously, at ground level, as the treetops a hundred feet in the air. And each for its size is doing a good job of plowing and harrowing and cultivating. Lower your eyes still farther and even the dry and apparently dead grass is responding to the tumult. The wind is pulling and tweaking and plucking at every strand and shred of vegetation. It is all part of the agriculture practised by nature in a far more subtle fashion than man has ever been able to devise.

A Measure of Worship

At the funeral of George Moore, the Irish novelist, the famous poet George Russell said: "To the artist, the love of earth, rock, sea and sky is a form of worship."

But artists are not the only characters who worship through nature. Many a good church-goer finds himself closer to the divine in his garden or in the fields or woods than at any other time. And perhaps the majority of those who don't go to church commonly experience some measure of worship and prayer in their love of earth, rock, sea and sky.

There is a rapidly growing cult of worshippers – they would be the last to admit the title – all over Canada and the United States. They are the bird watchers. With field glasses hung around their necks, a field guide to birds in their pockets, in old outdoor clothing and sturdy boots, they are ranging the parks and suburbs of cities, motoring and tramping the byways of the country, with no other idea in their heads but to see and identify as many different kinds of birds as possible in a day. It sounds a little absurd. But it happens to be one of the most exciting forms of hunting ever conceived by that hunting creature, man. It entails more outdoor exercise than any deer hunter or golfer or follower of the hounds can hope for. It is an all-season sport. Spring, summer, autumn, winter, the bird watchers are afield. They think more of a score of twenty birds seen in December than of a hundred seen on one May day. They kill nothing. They harm nothing.

Their numbers increase. They are a non-secret society.

In September, a mocking bird, a very rare visitor from the south, strayed into my garden. I telephoned a bank manager who is a bird watcher, though it was only 2 p.m. In twenty minutes, he had spread the word, and here's what was gathered in my garden: two bank managers, a furrier, a professional ornithologist from the university, a man who feeds the animals in a nearby park, a professor of history and a locomotive engineer.

Life List— A Lifetime Pursuit

Roderick Haig-Brown, the Canadian novelist and writer on the out-of-doors and country life, in his book *Fisherman's Spring* calls one of the chapters "Recognizing Birds." This highly intelligent title is a far better name for the hobby than "bird watching." For most of those who follow the sport of walking in the woods and along the country roads looking for birds do not really "watch" birds.

Most bird watchers merely find them, identify them and add them to their list for the day, or, if they are specially lucky, to their life list.

The life list, of course, is the basic incentive to most

amateur bird watchers. As beginners, they start a list of all the familiar species, the robins, the thrushes, the woodpeckers, learning to identify them unfailingly by sight and sound, and being increasingly amazed at the ever-mounting numbers of these hitherto unseen, unnoticed creatures of nature. By the time the life list of birds seen, learned, and definitely known on sight reaches a hundred, you have a real hunter. And prospecting for uranium has no more grip on a prospector than watching for a cerulean warbler has on a bird spotter who has never seen a cerulean warbler. And I'm one of them.

Land Rights

By the latter half of April the crows, two by two now, have arrived and taken up their territories. From ocean to ocean, on every acre of land, the native crows, millions of them, have come back from vacation, and have staked out their property for the coming season.

They arrived in companies and congregations soon after the snow went. Maybe a few mass meetings were held in woodlots for a few days. But for the most part, it was no time until they were paired off, and, with little visible dispute, had distributed themselves peaceably around each area until there was, you might say, not an acre of land untenanted or unclaimed. Across the whole vast face of Canada the crows are in possession.

We are in the habit of thinking that Canada belongs to us humans. With legislation and much surveying and engineering and much filing of deeds in registry offices, we lay claim to land. But the crows were here before us by nobody knows how many tens or hundreds of thousands of years. And they go right on being here, regardless of us. And the queer thought cannot help but intrude itself that they will go right on being here, even if we are not.

Long-Distance Passage

Any day late in April you may see in your backyard a strange bird, maybe gay, maybe quite drab; but strange. If nothing else attracts your interest in it, think for a minute of where it has just come from. And how.

At 25 miles per hour, it may have just come from Brazil. Flying only at night, with nothing to guide it but blind instinct, and resting and feeding by day, last week in Kentucky, the week before in Texas, and the week before that in Panama, this casual visitor to your garden may have come to you straight from Peru.

The first nighthawk you hear tweenting in the evening sky, consider him. He is a slow erratic traveller, feeding as he goes. But he happens to be one of the most contemptuous of all creatures of this solid round earth and the majority of its distances. The nighthawks of Alaska migrate, circling and tweenting, 7,000 miles to the Argentine for their winter. And now come all the way back. The one over your head has just done 6,000. He may nest on your roof.

Nineteen species of water birds nest north of the Arctic Circle. If you live near a lake, you will be able to see some of these quietly resting along the beaches. They are worth a glance from you. Because you are seeing, possibly, one of the six of these nineteen species which travels from the Arctic to the Antarctic Circle twice each year, 8,000 miles south in autumn, 8,000 miles north now in spring.

They think nothing of it. They weigh less than your hand. When they go south that 8,000 miles, most of them are not four months old.

From mid-April until mid-May, thousands upon thousands of birds will come pelting past you, over your head, in the night. Maybe 10,000 will pass within a couple of miles of you. From the wild goose and swan down to mites no bigger than your thumb, the travellers are hurrying home.

Mystery of Migration

We can guide an unmanned spaceship to Mars, but we still don't know how a bird, born in June on the tundra of the far Canadian north, with no older birds accompanying it, flies unerringly southeast across Labrador to Nova Scotia, takes off across the vast Atlantic and, ignoring any islands it may pass en route, reaches South America's coast, and thence proceeds, by the identical route its ancestors have followed for centuries, down to Patagonia.

This is what the golden plover does. And in some such style, all birds do it.

The warblers will soon begin to appear, anywhere from thirty to forty different species of them according to where you live, fresh home from Yucatan, Brazil, the Bahamas, Cuba, Colombia, Nicaragua, Venezuela. They went there at four months of age, and most of them aren't a year old yet. But the distances, the storms, the nights, the perils of every imaginable sort were all borne by a body weighing less than a rosebud, and a tiny brain which contains some sort of mechanism that so far defies all enquiry of the scientific mind that has fissioned the atom.

It is a mechanism that controls the birds as wholly as remote control guides a rocket. But what or where that mechanism is in the bird's brain or nervous system, science has so far, despite tireless study during the past enlightened half century, been unable to discover. Isn't it odd that the eggheads who gave us the hydrogen bomb are not more interested in Who handles the remote control of birds?

Arrivals on a South Wind

Starting in the first ten days of May, every south wind, whether with a westerly or an easterly bias, wafts literally millions of birds over the border and into our midst. The day following a southerly breeze, a stroll in thicket and field will be in the company of pilgrims full of that spirit of ecstasy that characterizes pilgrims anywhere who have come thousands of miles to their shrine. Here is their shrine.

Many of the birds have been back for weeks, the hermit thrush, robin, red-wing and many of the sparrows; but now come cascading in from the south the warblers, the indigo bunting, rose-breasted grosbeak, the tanagers. With them, they bring the first music of any volume. The morning after the next southerly wind will be the occasion of the opening symphony concert of the year.

These are the birds that have come the farthest. Among them, the tanagers offer the weirdest problem. The western tanager winters from Mexico to Costa Rica. The scarlet tanager, familiar in the east, winters from Bolivia to Peru deep into South America. They are characteristic tropical birds. In the tropics, and never leaving the tropics, are some hundred and more different species of tanagers. Why do these three tropical birds, the scarlet, the summer and the western tanager, leave Peru and Bolivia and, after crossing the tropics, head thousands of miles north to nest in Canada? Why does not one of their numerous close relatives come? Do any of these travelling members of the large family remain at this time of the year in the tropics? Apparently not. These tanagers are worthy of a respectful stare whenever you see one.

City Migrants

By a curious coincidence, I went to live in a house built on a ravine that, fifty years ago, was far out in the country, beyond the suburbs of the city, and with which I was very familiar as a boy searching for birds and wild flowers.

A brook ran down this ravine, and it was a veritable spring highway for the migrating birds. We used to come and explore the creek, building camp fires to cook lunch and rejoicing in the wilderness.

But the city grew out and engulfed the field and ravines. This ravine was filled in solid, and the creek that ran down it was diverted into the drains. Twenty-five years ago this house was built, and then the city reached miles beyond.

Yet, to this day, though there remains no vestige of the ravine, all being flat expanses of concrete and brick alive with the traffic and movement of humanity, the migrating birds come full force as they did fifty years ago, following the ancient route across the miles of city, and through my garden as if the ravine and the brook were still here. Not the odd bird, but dozens and dozens, of all the species I knew as a boy here, and in about the same numbers. Imagine a woodcock springing up under your feet in a backyard in the heart of a city? The garden was a ravine and a brook for five or ten thousand years. After the woodcock come the hermit thrushes and the fox sparrows, then the white throats. A whip-por-will on the flat scantling of the grapevine. The warblers come tumbling through, perhaps six or seven species all in one morning.

These birds follow some mysterious paths unknown to human intelligence. Or, come to think of it, do they expect my house and my city to vanish presently, and the brook and ravine to return?

Warblers

All across Canada, in the first week of May, the warblers have commenced to arrive. These are the absolute jewels of the bird world. In most parts of the country the migration reaches its peak about the 17th of May. Around that date, you might be lucky enough to see twenty or more different species of these incomparable atoms of nature on one perfect afternoon.

There are fifty-five species of warblers in North America, but the men who have seen them all are mighty few and far between. In Ontario, for example, fewer than forty species have been reported over the years, and to see that full number, even in a lifetime, an amateur naturalist has to be a fanatic. I am rather stuck-up over having seen twenty-six species in fifty years.

Through field glasses, many of them – the blackburnian, the chestnut-sided, the parula – are fabulous to behold. They are not much larger than your thumb. They gleam and glitter with vivid and pure colour, predominantly yellows and oranges, on soft greens, with spectacular markings in black or white. Though they are called warblers, their songs are small, sibilant buzzings or brief series of briefer notes, zipped rather than whistled.

The place to see them is preferably on a wooded hillside along a creek or pond. Parks in or on the edge of cities are often as good places as in the country, for the warblers are on the move, heading north, and their swarms take no account of cities in their path.

On a soft, sunny day, sit down on the hillside where you have a fair view amid the upper branches and tops of the trees and simply wait for these jewels of the Madonna to pass by. Once they have passed and gone to their nesting regions, you might have to travel two thousand miles to see what you may see in an afternoon, sitting quietly, amidst the migration.

Antic Courtship

No two species of ducks follow the same ritual of courtship. But each observes its own ritual as if it were laid down in some sort of duck Hoyle, and can be observed while the wild fowl are choosing mates in preparation for their northern journey. In early spring, wherever you can find open water on stream or lake, your binoculars are likely to find ducks putting on their ceremonial.

As with humans, gay clothing seems to be an important feature of courtship. A beautifully coloured duck like the male American merganser, a big saw-bill or fish-duck, goes through a most energetic display of all his raiment. He swells up his breast to show off the delicate peach or salmon-pink suffusion of it; he rears up on his tail to exhibit his magnificent white flanks, arching his neck to show off his glittering green head and red beak. And then, as if that were not enough, he whacks down in order to kick a spurt of water three feet in the air, only to expose his beautiful red feet. He is a complete show-off.

The whistler or golden eye, on the other hand, being mostly black and white with an irridescent green head and vulgar yellow feet, goes in for contortions to show his lady what a character he is. He throws his head back, beak vertical, and utters a wheeze like a night-hawk; then jerks his back almost onto his rump; and as suddenly jerks it forward, as if he were trying to break his own neck. What shenanigans, all for love!

A Billion Voices Singing

Most of us get out into the country at the stupidest time of the year when, in the sultry summer, all nature prefers to sleep in the shade – except dizzy, noisy, holiday-making humans.

The birds, by summer, have nearly all ceased singing. But from the May holiday to early June, the enormous sky across Canada, from Newfoundland to the Queen Charlotte Islands, rings with their up-pointed music. It takes a little imagination to try to conceive the amount of sound the birds are making at this moment. The National Audubon Society submits that there are 15,000 million birds in North America at any given time. Of this number, we may have half or better in Canada, and all the males are whooping it up.

Loons are yelling, owls are hooting, gulls over their nesting rocks are yodeling and screaming. Such a din is there across the Arctic and subarctic tundra, as the millions of waterfowl and shore birds nesting there give voice to their feelings, that northern men tell me the racket is enough to drive you bushed. In the eastern forests, on the prairies and the mountains, not a single acre of Canada but has from two or three to twenty nests of various birds; and over each, a bursting gentleman uses the vocal instrument God gave him with all his force.

Most of them are worth hearing. A Tibetan monk with his fifteen-foot horn can't make, in proportion, as terrific a sound as a loon can with its careless throat. And after you have listened closely to a hermit thrush in the evening treetop for half an hour, you will have to confess that human ingenuity has so far failed to produce any instrument, any violin or flute, to equal for sheer beauty of sound, the trivial little mesh of muscles in this small bird's neck.

Downtown Birds

The house sparrow, the starling and the common pigeon – all European – are about the only birds you will see downtown in cities. Rarely do you see native birds venturing so far. Robins will nest in mid-town areas that have parks or large lawns. In Montreal, for years a peregrine falcon, or duck hawk as it is more commonly called, nested on the noble pinnacle of the Sun Life building. In Toronto, the smaller sparrow hawk has nested in the steeple of Upper Canada College, and has been known to take up residence on sky scrapers. Such events are so rare as to warrant news items in the press.

The imported birds, sparrow, starling and pigeon, normally build around houses. But so does the native phoebe, which scorns cities, especially downtown areas, perhaps because there are no flies of any account to feed on.

The minute the downtown area begins to give way to gardens, however small or dingy, the native birds start to occur. In eastern Canada, you can find catbirds nesting in the ornamental shrubbery of a pumping station, a chipping sparrow in the spiraea in front of a crowded school; hummingbirds are not at all uncommon when and bigger gardens begin. And it is the less tidy gardens that attract most birds.

By the time you get far enough out of the downtown to find parks, the native birds commence to chum up to us. Owls, woodpeckers, several kinds of native sparrows, all better singers than canaries, and a few finches, warblers and vireos will raise broods in parks through and around which snorts our maximum traffic.

One of the first things a newcomer from the Old Country comments upon in Canada is the absence of wild birds around the town. The sparrow, starling and pigeons are chummy enough. It may take our native species a century or two to warm up to us.

Nest Robbers

As May advances, you will see the crows, jays and cowbirds doing their best skulking. Crows and jays are, for about forty-eight weeks of the year, the rowdiest and noisiest of birds. But in May, all of a sudden they grow furtive. They become sneaky and incredibly sly. Without a sound, and with none of that arrogant flight from treetop to treetop, but coasting, as it were, low down amid and among the trees and bushes, the crows and jays are playing Indian. They are looking for birds' nests to rob. They want to take home to their own nestlings something nourishing and warm.

Without the slightest sound, a crow will suddenly zoom up in the midst of a glade, and take an inconspicuous perch in a dense tree. There he will sit, peering, listening. In short, looping flights he will move cautiously through the glade, as quietly in broad daylight as any owl at midnight. What he can't see he can hear. For nestling birds almost invariably give themselves away by some small, raucous or hissing clamour when the parent birds arrive with food for them. That buzzing the crow listens for. You see him relaxing comfortably, then moving towards the vireo's or the thrush's nest to take his pickings for his own hungry and gape-mouthed young.

The jay plays the same game. That raspy, shouting fellow becomes mute with parental solicitude.

The cowbird is the champion sneak of them all. As long as any nest is still abuilding, Mr. Cowbird in on the make. How he sits and peers. Mrs. Cowbird, a nasty little woman, waddles and wiggles about in the grass and shrubbery, searching for a nest, any nest, into which she can unload an egg. She is dead grey, curiously featureless, unlovely – the fallen woman amongst birds.

Airt

A problem that engages the minds of countless thousands of people from April to October is: what is a good fishing day? Some swear by the old rhyme that dates back away beyond Queen Elizabeth I:

Wind in the north,
  Wise fisher goes not forth.
Wind in the east,
  Fit for neither man nor beast.
Wind in the west,
  The fishing's best.
But wind in the south
  Blows the bait in the fish's mouth.

Nearly everyone agrees that a dull, humid day, with rain either threatening or first falling, is the perfect fishing day. Yet everyone recalls having some of the greatest catches of his life on brilliant hard days with a spanking wind blowing. In east winds, with a cold rain blowing horizontal, wonderful baskets of trout or bass have been taken, to set at naught all the rules and to make more mysterious than ever the question: what is a good fishing day?

In Scotland, the gillies have a word that defines a good fishing day. If a day has an airt, it is a good day for fishing. What an airt is, no dictionary defines; and no two Scots will agree. When a gillie proceeds to find out if a day has an airt, he takes a pinch of air in finger and thumb. He feels the texture of the atmosphere in his rough and horny hand. There is something psychic about it, too. The gillie, while fingering the air, glances about at the landscape, at the sky, and then assumes a trance-like expression. At the same time, he may breathe deeply, run his tongue over his lips, give a couple of good Scottish sniffs and then say:

"Oo, aye, there's an airt the day!"

And when he says so, the fishing is bound to be good. In Canada, we do not have airts. I have seen Scotsmen transplanted here going through all the ritual, fingering the air, assuming the trance, and all. But the airt eludes them. Yet they are on the right track. For there is something mysterious and psychic about a good fishing day that defies definition. A good fishing day is a day when the fishing is good.

Trembling Aspen

The trembling aspen is best known to Canadians as the common poplar. The legend of olden time is that it was from the poplar that the cross was made on which Jesus died, and for remembrance, its leaves have trembled ever since. The slightest breath of air agitates the whole tree.

It is an all-Canadian tree. Except for a small area of southern Alberta and Saskatchewan, and the coastal strip of British Columbia and Vancouver Island, it covers the whole realm from Newfoundland to Alaska and right up to the tundra of the Arctic.

Most people have not much of an opinion of the poplar because of its comparatively brief life, its soft wood, its greedy weed-like habit of seizing every opportunity to spring up and flourish on soil we might hope more profitable trees would inherit.

But Canada owes an enormous debt to the poplar. For it is the greatest of the resurrectionists. When forest fire has laid waste vast expanses of the wilds, it is the poplar that first comes to restore the blessed green. Like all aspens, this trembling one reproduces quickly by root suckers that escaped the devastation.

First comes the poplar to make a windbreak and a shelter for the seeds of spruce and pine and vegetation blown on the winter gales from far away. And by the time it has lived its thirty, forty trembling years, it has made the coverlet for the more enduring forest that is to come.

Canadians should look friendly upon the trembling one.

The Glimpse

Somebody might be interested in a conversation I had with an embittered old gentleman at the wonderful boat exhibition during a sportsman's show in Toronto.

"I dunno," he said, "how many hundred boats there are on display here. Hundreds and hundreds, anyway. So far, I have seen only eleven canoes and three rowboats. All the rest are yachts, cruisers, speedboats and literally swarms of little teeny weeny skiffs with great big outboard engines perched on their sterns. I bet there is hardly a boat in this whole show that can't go twenty miles an hour. Most of them can go faster.

"Well, sir," said the old-timer ruefully, "I mind the time when close to one hundred per cent of the summer cottagers and holidayers of this country owned a rowboat, and nothing else. Maybe ten per cent of them had canoes, though the canoe-owners were considered pretty venturesome. Rowboats were the normal and natural means of transportation for all and sundry.

"Why? Because the going was just as important as the getting there. People wanted to see things, in those days. They wanted to take in the scene. But I want to know something about these folks nowadays. What can you see, going licketty-split, in a cloud of spray? Isn't anybody interested in looking any more? Or is it that all they want out of life is a glimpse?"

I shook the old gentleman's hand warmly.

Talisman

Most of us carry, like talismans suspended in the secret and precious back rooms of our minds, one or two or three memories of something seen which, in an instant, can be seen again, as clearly and poignantly as ever. They are fadeless.

Wild nature supplies an immense number of us with these memories. Sunsets, storms on water, a flight of wild geese, a first glimpse of a landscape, an aurora. We do not have to be nature lovers or even particularly attracted to the natural world to have these etchings imprinted on our memories, in an instant, and to last forever. Two of mine that recur endlessly, without being summoned, may spring into my mind's eye while I am shaving, or walking in a crowded street. The first vision is that of a grey, lifeless November day, near sundown. I was deer hunting, and came out on a high bald rock in the Cloche Mountains. No wind was blowing. The landscape before me was lifeless and still. Below me was a wide gulley in which a swampy creek ran. As I stood watching, there appeared far down the gulley a wispy, silvery flight of small birds. It was the advance guard of an immense flock of snow buntings. They did not pass in a solid flock but in a silent, undulent stream, uttering no sound whatever, below me, so that I could look down on their white and buff colouring, a ghostly, perhaps weary procession of hundreds upon hundreds of them, dipping, rising, dreamlike, magical.

The other talisman is my first sight of a showy orchis. I had hunted for twenty years for this orchid; all my sporting companions and most of my friends and family had seen them. But by one of those complexities that beset those who love the outdoors, I just could not find one.

Then one spring day I was searching for an oven bird's nest. I had tracked the singing bird down. I knew the nest was within a limited area. I was down on hands and knees intently scrutinizing every foot of the ground. And there, under the beeches, two feet from my eyes, was a showy orchis in full and fabulous bloom. It was the least showy of all flowers. It was like a tiny nun, white and mauve, shrinking, cloistered. It will live there, in that place, in that instant, as long as I live.

Feather Trade

Somewhere around 1735, when George Washington and Bonnie Prince Charlie were both boys, a few fishing boat owners along the New England coast decided to go into the feather business. Their fishing crews that went up to the Newfoundland Banks knew all about the vast, the literally countless flocks or herds of ducks that nested on the islands of the St. Lawrence Gulf, along Labrador and around Newfoundland. These fishermen, like their predecessors in the trade right back to Jacques Cartier, used to depend on these wildfowl colonies for a good deal of their provisions while at sea fishing.

All the sailors had bunk ticks of down. They used to bring home bales of eiderdown for domestic use. So the businessmen of New England organized the business. They outfitted ships expressly to go north, in spring, to these nesting grounds of the countless multitudes of wildfowl and to bring home shiploads of feathers. Not only America, but Europe and Asia, developed rapidly into customers of the trade. The whole world slept in feathers.

The competing ships made short work of the wild treasure. The crews would invade an island, encircle the ducks, flightless at this season, drive them into a huddle and slaughter them not by the thousands, but by the hundreds of thousands. It is all on the record. In twenty years, between 1740 and 1760, they had exterminated the Labrador duck, a few of which lingered on for nearly a century, the last Labrador duck falling to a gunner near New York City in 1875. But the various eiders and other wildfowl that had been one of nature's massive glories, like the buffalo and the redwood forests, were reduced to a shadow.

All in twenty years. All by a few companies of business adventurers. Over and over again, the story of man's ravenous talent for exploitation has been and is being demonstrated. But nothing quite equals in ferocity the feather trade of 1740-1760.

Nature's Health

The biologists studying our fish and wildlife resources are coming to the opinion that good years and bad years have existed for countless ages before man started to upset the balance, and that the real reason why fish, say, are scarce this season in certain waters is due more to a bad spawning season four or five years ago than to the activities of even a swarm of fishermen.

This does not mean that the presence of man has not something to do with the scarcity. It means that the presence of man, with his dams raising and lowering water levels, cutting off the timber, clearing the land for cultivation of farms, filling the water with oil and acids from his highways, engines, boats, and otherwise interfering with the natural resources of the fish, has upset the ideal spawning conditions that make for a plentiful supply of fish.

A thousand years ago, any given lake you may have in mind used to have good and bad years. Cold, changeable weather, for six years in succession, could almost wipe out the population of any fish. Then, following two or three years of weather ideal for spawning, the lake would fairly jump with fish.

What the biologists are saying is that we should take as much account of the public health of nature as we do of the public health of us two-legged consumers of nature.

Everyday Song

After the dishes had been swished clean in the sandy water and the frying pan swabbed dry with bunched grass, the Indian sat down on the far side of the fire, drew his knees up, rested his arms on them and began quietly to sing.

In the dusk, it was a curious sound. It was more like monotonous humming: but I could detect words. The Indian's eyes were gazing absently at the ground.

"What is that?" I asked him.

"I am singing my everyday song," the Indian replied.

"Everyday?"

"Yes," said the Indian. "I am praying."

Because we had become friends that day and caught not many but very large trout, and because I had listened respectfully to him for a long time while he told me about his son who had died of polio (instead of interrupting him to tell him to get on with his paddle) the Indian smiled, turned his head to look at the ground again, sang a little louder and translated his song into English.

"I wake," he intoned, "and the mist is on the water. I open the tent and see Shongwesh – that is, mink – in the sport's fish basket. I clap my hands and he runs to the water. I look at the sport. He is awake and winks his eye. I walk down to the water. To wash my hands and face. The water is cold. I splash water onto the long grass and Shongwesh runs out. He was hiding. When I stand up, the sport is sitting up in his bed lighting a cigarette. I walk to him and say two eggs or some ham or trout? . . ."

In a murmuring sing-song voice, the Indian recounted every single, solitary moment of the day that was ending – the fishing, the paddling, the points rounded, the beaches landed on. It took nearly an hour.

It is the Indian's prayer; and a marvellous way to thank God for a day.

Victims of Folklore

Certain animals and birds have been given a bad reputation by legend and literature which, on reasonable examination, does not stand up. The wolf and the raven are two handy examples.

The wolf has come down through folklore and into classic and modern literature as a fearsome and ravening beast.

I helped Andy Tyson, a trapper and woodsman well known in the Sudbury and Manitoulin region, to take a magnificent hundred-pound male timber wolf dead from a snare.

It would be hard to imagine a more superb animal. Its head, immensely broad across the brow, was more beautiful than that of any breed of dog. Its coat was rich and healthy; limbs powerful and lithe; paws broad and in wonderful condition. As I stood looking at him, I had the queerest impulse to take off my hat in reverence. What generations must lie back of him, across the cold, wild ages of this country!

We hate him. We put a bounty on him. For centuries we have hunted him with trap, gun, snare, poison. We hate him because he kills the deer we want to kill; he steals the fowls we intend for our pots. But principally we hate him because we are frightened of him. Even lying there dead before us, he was not beaten. We had merely tricked him. . . .

Overhead, eleven ravens flew. Andy Tyson smiled up at them: "Mating," he said. At first glance, they look a little like crows. At first glance, a penned turkey might look like an eagle. Folklore and literature hand ravens down to us as birds of ill omen. They are beautiful, soaring, rugged birds of the lonely places. They are not furtive, ragged birds like crows, dangling like old umbrellas across the wind. The ravens, muttering and croaking overhead, had to my mind something of the indefinable nobility of this platinum grey wolf on the ice before us, stiff, stark, unbeaten, tricked.

Emancipated Guides

Tourists and sportsmen are finding out to their dismay that in a good many parts of Canada the guides work on what might be called union hours, though they don't belong to an accredited union. Or let us say the guides work on "modern" hours. That is, they quit in time to be back at hotel or resort by 6 p.m. And of course the best fishing of the day is often between 6 p.m. and dark.

A few years back, guides were a class of men apart in this materialistic age and country. They belonged, body and britches, like a serving man or even a slave, to their master for the day at $5. They appeared at daybreak, ready to go long before the sportsman. They toiled all day like old-time stevedores, paddling, portaging, carrying fantastic loads, preparing meals, demonstrating the shrewdest knowledge of where the fish or game were, and on top of all that, showing considerable talent as mind-readers, anticipating every wish or whim of their masters. Such guides are still to be found in some of the less frequented regions of Canada.

But the modern guide on duty at a great many of our liveliest tourist resort areas is an emancipated man. His wages are more than double the former rate, with expectation of tips added if he delivers the goods. And he quits in time to be home for supper. He is, in other words, hep to his age and generation.

This is not a bad thing. For the sake of conservation of our wild resources such as game fish and game in season, I like to think of all those swarms of tourists, blessed though they may be for contributing to our economy, having to quit fishing just about the time the fishing gets good. A wise old fish, of the size and quality to be stuffed and hung over a fireplace in a far distant rumpus room, lies doggo all day in some deep hole, under some sheltering ledge, and does not come forth to feed until dusk approaches and all the guides have shepherded the anglers home. It is towards dusk the deer or the partridge quits his holt to start roaming in the gloaming.

It is only right that in view of the enormous increase in the ease of sportsmanship, such as the ease of transportation with outboard motors and aircraft, and by motor roads into the vastness of the north, something should make it harder to kill the game. And the guides, emancipated, are providing the handicap.

SUMMER

Discrimination Against Cats

Mid-June is a bad time for cat owners. This is when their neighbours suggest, more or less amiably, that they either tie a bell around the cat's neck, or else keep the blame thing in the house.

For now, the young birds are tumbling from the nests all over the city gardens and parks. And on lawn and fence post, they are feebly and yet violently advertising themselves to all cat creation.

One of my neighbours is the type of city man who, though he wants a drug store no more than two blocks distant, likes to preserve a little oasis of the country at the back of his house. And he has hedges and apple trees to wall out the city from his view. His garden is rustic. And he loves to entice wild birds to nest, with boxes and shelves cunningly hidden amidst his greenery.

He is definitely anti-feline. He is guilty of the grossest discriminations against cats, even to keeping a pile of things on the garden table to throw at them if they come near his sanctuary. He is an ardent cat-bell man, and even buys cat bells to give to his neighbours who own felines.

But he does love a cocker spaniel which he has trained to do everything but leave baby birds alone. I am sorry to report that one of his cat-owning neighbours, having detected the cocker in the act of scoffing off a baby oriole, sent my friend a cow bell on a nice big blue ribbon.

The fact is, wild birds should be discouraged from nesting in cities. Besides cats and dogs, they have traffic to contend with. And more infant birds are probably killed by garden insecticides than by all the cats in Christendom.

Nature sometimes appears to be very foolish. You will often hear, in your garden, a curiously low, furtive robin call, a single note, uttered at intervals of about five seconds. It is the warning note of a robin to its nestlings to keep still, or to its fledglings on the ground to watch out. For a cat is about.

When you hear that low, mellow single note, you can be perfectly sure to discover a cat in the immediate vicinity. In fact, you may find two or three cats. Because if there is anything calculated better to bring a cat in a hurry, it is that robin note advertising all over a whole city block that fledglings are on the ground somewhere within a few yards of the adult bird. Any intelligent cat can have robin chick for supper merely by scouting a very small area around the parent.

Many other birds have alarm notes and warning calls, some of them so loud and strident as to be heard over a quarter of a mile, and, like hunters' horns, spreading the news far and wide that game is afoot. Every predator within hearing, the hawks, crows, weasels, prick up their ears at the sound, and come to the dinner call. You would suppose that nature, which gave the partridge, for instance, the instinct to feign injury to seduce enemies away from her nest, would have worked out across the ages a less fatal system than the broadcast outcry of many of the most delicate creatures.

Nature, however, is seldom foolish. For we are wrong to suppose that nature is interested solely in robins and other attractive species. There is no reason whatever to believe that nature is not just as interested in weasels, hawks, and crows, as she is in robins, bluebirds and sandpipers. And when you hear a robin uttering its fatuous warning call, it is merely nature thinking kindly of cats.

Nature's Laws

There is a sort of eager obedience in nature. Everything in nature, save man, toes the mark.

In the woods, the first act of spring, in eastern and central Canada, is the flowering of the trilliums and dog-tooth violets – or fawn lilies, as they are better called. In the west, closely related species serve to raise the curtain.

The carpet of the woods consists of apparently little else than these two or three plants. They possess the soil. You walk ankle deep through the fresh, rich foliage of trillium and fawn lily, countless thousands of them, all burdened with their flowers.

A few weeks later, when June comes, you walk in the same woods, and you have to search and peer to see even a few of the now flowerless, retired trilliums and fawn lilies. They have had their turn. They are obedient. They have withdrawn from the scene, under new and abundant plants that were not to be seen at all while the first-comers had the stage. Their leaves seem to have shrunk in size. The plants appear to be ever so much fewer in number.

Some of the philosophers of nature stress the struggle of nature, offering us the picture of a pitiless struggle between species that smother one another, eat one another in an endless war of survival. If some species do, they are merely obeying laws. But in nature also is this synthesis or processional ritual, in which one creature, plant or animal follows another, with time allowed for all – a perfect program. If they eat one another, it is for the common good. If they wantonly destroy one another, it is for the common good. Man, of course, is different. He plans to let nothing eat except those things he proposes to eat himself.

Mudcat Mystery

How in the world, a friend of mine wants to know, do the catfish get into all the farm ponds all across the country? A farmer gouges out an excavation to make a pond for his cattle. No creeks run into it, or out of it. Maybe a couple of springs feed it for a little while in spring and fall. Presently the cat-tails begin to fringe its margins, and, miraculously, pond weeds of one kind and another begin to flourish in it.

And then the catfish, the little mudcats, appear.

A biologist offers this explanation. The blue heron knows where all the ponds are, as he waves his solemn way across the country. He drops in on a lake or pond where the catfish properly belongs. He stalks along the muddy shore, where the mamma catfish is herding her swarm of infants. The heron steps into one of the herds with his mud-caked feet and legs. Catfish are incredibly recuperative. They will survive a long time in the fisherman's boat. The heron, with the infant catfish stuck in the mud of its feet, flies off to the next pond.

Or maybe it is fertile eggs of the catfish that the heron carries, as it does the seeds of aquatic plants, insects and many such initial transplantings of life. Ingenious nature gives herons big feet.

Martins

It is never the wrong time of year to build a martin house on your property, and if you build it during the summer, the chances of travelling or migrating martins seeing it and coming back to it next year are good. Some people have put up elaborate eight and sixteen apartment martin houses and waited five or six years before the birds adopted it. But once they come, they come apparently forever.

And no sweeter, homelier sound in nature exists than the day-long musical chatter of these sociable big swallows. Except in the interior of British Columbia, the martin is local and often common across the whole southern belt of Canada.

On a remote and fairly barren section of the Great Lakes country, a cottager had had a martin house for twenty years. Then one winter, a storm blew it off its fourteen-foot mast and destroyed it. Early in May, the cottager went up in the hope of getting it rebuilt before the martins returned, found it gone and departed elsewhere. The local report was that the martins had already been there two weeks. Apparently they had gone. None was to be seen around the area.

The cottager spent a morning and part of the afternoon building an eight-apartment house, and erected it back on the fourteen-foot pedestal and mast, solid and secure. He stepped back a few yards to see if it was plumb. And as he did so, three male martins flew onto the house and with wild chirping and chuckling proceeded to explore the new apartments. Before supper, there were eight male martins in possession. By nightfall, eight females had arrived out of the blue. When darkness fell, eight pairs were cosily and peacefully inside the eight tiny rooms.

Lady Slippers

The small plane in which I was a passenger on a fishing trip was forced down with engine trouble in a wilderness lake many miles off the beaten track, even of aircraft. While waiting for the other plane which we summoned to bring repair parts, I took the opportunity to do a little exploring in country that almost certainly had very seldom had any human beings in it, even Indians. No doubt in winter a lone trapper might penetrate in there; but even that was doubtful, so detached was the lake from streams, rivers or chains of lakes which might have invited approach.

I went out onto a heavily wooded point of land and found it to be literally carpetted with perhaps the loveliest of all Canadian orchids, the lady slipper that is called either showy or queen lady slipper. There were thousands of blooms. Amid the cedars, spruces and birches, the glorious abundance of the flowers was breathtaking. Their colour is pure snow-white flushed with magenta; and the green of their foliage is of a liveliness befitting the exotic splendour of the blossoms.

It was while staring at these acres and acres of lavish beauty that I detected in my thoughts a curious indignation. Except for an accident that brought us down here this wonderful display would never have been seen by human eye. For whom was all this natural loveliness designed? It was not just a chance that it was here. It had been here, unseen by any eye save vagrant birds, for years, probably for centuries past. And for whose edification?

In a way, it was a laughable reaction I was experiencing. Yet also it was tragic. The vanity of the human spirit is such that it supposes all nature's glory is for human consumption. The truth must be that nature conceives her beauty with an intention larger than we can perceive. Mere bees is not answer enough for acres of the queen lady slipper.

The Raccoon – Predatory, Personable

Naturalists are still undecided as to which animal or bird is the greatest predator at the present time. But the raccoon is steadily gaining converts. A fox can't climb trees. A fox does not too cheerfully go in swimming in order to reach a duck's nest in a stump fifty feet from shore. But the raccoon can pretty nearly reach anything he takes a fancy to in his nightly hunting prowls. And his fancy is free.

Only the fact that he consumes prodigious quantities of insects of every sort allows him into polite society among naturalists. He lives near water, as a rule. That is why he is little known on the prairies. And his handiest food he finds along streams and lakes: frogs, crawfish, minnows, snails, and a tasty turtle's nest full of eggs. But living in hollow trees for preference and being as nimble as a squirrel in trees, he hunts birds' eggs and nestlings diligently, from warblers' to owls' and plays havoc with ducks' nests, since he is as much at home in the water as he is in the trees. His forepaws are literally hands. He can reach into holes and fetch out a woodpecker's or a bluebird's eggs as neatly as could a monkey. He is a great fighter, of course, as any hound man can tell you, and is able therefore to eat a wide variety of birds and animals. Plenty of mice go into his diet. But, being a member of the bear family, he is also a great berry eater, and fond of many a vegetable titbit, the favourite being a farmer's choicest corn on the cob.

His scientific name is one of the loveliest in the books – Procyon lotor. It means one who snarls like a dog and washes. The coon loves to squat by the waterside and swish his food about in the water, washing it of grit, and manipulating it in his nimble black hands with the loving dexterity of a card sharp shuffling the cards. With his black mask over his eyes, his prison-striped tail and all his wicked habits, the coon is the scamp of the woods; and a great pleasure to know.

Shearwater

Everyone who has travelled on the Atlantic will recall the shearwaters, smaller than gulls, with stiff wings, sailing low over the waves, curving down into the troughs, greyish birds on top, white below, as plentiful in mid-Atlantic as they are a day out from land. The life story of the shearwaters must be amongst the strangest in all nature. Off the coast of Wales is the island of Skokholm, where R. M. Lockley lives, devoting his life to the study of the Manx shearwater, which nests in thousands on the island. Reviewing Lockley's book on shearwaters, *The Auk,* the American ornithological review says:

"Because of their enemies, the great black-backed gull, the shearwaters come to their burrows only during the hours of darkness. Apparently mates never see each other, but must recognize each other by voice, there being great variety in the screams and howls of different individuals. Parents never normally see their single chick, nor the chick them. Incubation lasts 51 days and fledgling 72 days. Each parent incubates for three to five days at a stretch while its partner is fishing far away. When the chick is about sixty days old, the parents cease their visits to it. The young bird stays in the burrow for about six days, and then comes out each night, for another seven days, to exercise its wings."

The night of its 72nd day, it makes for the sea, sometimes a long distance for a fledgling to scramble; sometimes daylight overtakes it, and the black-backed gulls.

But once in the sea, its strange and wonderful life of ranging hundreds and even thousands of miles across the vast deep, skimming like a swallow in tireless and endless flight, seemingly unaware of everything in the world, its fellow birds, the passing ships, the storms, the seasons, everything except the shape of each wave upon which it seems intent.

The homing instinct of the shearwater is legendary. Banded birds liberated in Venice, Switzerland and Boston arrived back in Skokholm as calmly as we might from a weekend fishing trip.

Pike (Great) (Northern)

The commonest fish in Canada, found from coast to coast, though rare in British Columbia, is the pike. It goes by a great variety of names – jack, jackfish, snake, grass pike. Canadians generally have held and still hold it in small esteem. Great numbers of Canadians won't eat it, though the odd gourmet asserts the pike's flesh is white, flakey and delicate. Those who are not gourmets feel they can taste mud and weeds in it, and class it with trash fish like the ling and the dogfish.

About ten or fifteen years ago, some of the American outdoor and sporting magazines began to dramatize the pike. In view of the rapid decline in the principal game fish of the continent, the trouts, bass, muskies, the tourist trade and the sporting goods industry, now capitalized at goodness knows how many billions of dollars, began to look with ardent respect upon such fish as crappies, perch, sunfish. And, of course, pike. We began to see dramatic cover paintings of huge pike curved in the air, exposing savage fangs and viciously shaking a lure from their jaws.

To call them pike would not do. In the sport magazines and on the air, the Americans began referring to them as Great Northern Pike, with capital letters. At first it was a little comic. True, in the real northern Canadian lakes, there were still to be caught huge pike and numbers of tackle-busters of ten pounds and over were fairly common. But down in the main tourist belt, where the average pike falling to the visitor's lure ran around two to four pounds, and pretty flabby at that, it struck Canadians as highly ludicrous to hear the guests referring to them dramatically as Great Northern Pike.

This is the century of the common man and of the

common fish. In only ten years, the beatification of the pike was complete. The sportsmen no longer refer to them as Great Northern Pike. They have cut "pike" off altogether, and with flashing eyes, they call them Great Northerns. "Been up in Canada," they say, "fishing for Great Northerns!"

Cuckoo Magic

A hundred years from now, our kin will probably know all about the mystery that involves the black-billed cuckoo. Right now, we are steeped in ignorance and superstition! I refer, of course, to our scientists as well as to Tom, Dick and Harry.

The cuckoo, a beautiful, furtive creature of tropical design and grace, is one of the few birds that will eat the fearfully destructive tent caterpillar. It is distributed all across Canada from coast to coast, but mostly along the southern edges of the provinces.

Near Matheson, in northern Ontario, there was an outbreak of the forest tent caterpillar, a species that builds no tent but defoliates the trees just as avidly as the common tent caterpillar we know in the southern regions.

Now, that district was between a hundred and a hundred and fifty miles beyond the previously known range of the cuckoo. But into the plagued area came the cuckoos in large numbers.

How did they know of the presence of the tent caterpillars? Does a sense of smell function over hundreds of miles? Some observers suggest that each species of birds have far-ranging members of their tribe who explore well beyond the normal range. Can birds talk to each other, then?

All we know, in our ignorance and superstition, is that when a plague of tent caterpillars breaks out, hundreds of miles away, the cuckoos come.

Seventeen-Year Locust

All over America, large numbers of country people still hold the superstition that when there are unusual numbers of seventeen-year locusts, there is going to be a war.

The seventeen-year locust is that sturdy insect that, in most parts of Canada, sits up in a windy tree and sings in a long, drawn-out, penetrating, high musical zing. It is also called the harvest fly and the cicada. On its wings is clearly featured the letter W. And in the American south, its eyes and head are suffused with a lustrous red.

The most arresting thing about this insect is the fact that it is actually seventeen years old when it sits up there in a breezy pine or oak and sings. For sixteen long years, that insect has been a grub down in the earth, subsisting on the juices from plant roots. Sixteen years is a long time to be a grub in order to enjoy a week or ten days of adult and winged freedom in the wide air. Uncles find this insect very useful when moralizing to young ten-year-old nephews. They point to the invisible creature, zinging so piercingly in the tree, and tell the boy that it is a bug six years older than the boy. That puts nephews in their place.

One of my nature-student friends reports a great sight. A blue jay, which is a pretty rowdy bird, caught a cicada and started to fly away with it. The agonized cicada started to buzz. Now, it must be fairly startling to be a blue jay and have, on the end of your slender beak, an insect that suddenly starts to buzz like a vice-president's buzzer.

The startled jay made an emergency landing on a limb, held the cicada down with its foot and examined it in consternation. The buzz ceased. The jay resumed the journey. The cicada burst into furious song again. The jay made another very bad emergency landing and firmly packed the offender to definite silence. The bug was possibly ten times older than the bird.

Vandal

The owners of a summer cottage had week-ended there, probably leaving an opened tin of jam or a package of bacon or some such treasure to reek out the key holes and door cracks for the nose of an exploring bear.

The bear had simply torn the screen, smashed a window pane, reached in and yanked the window, frame and all, out onto the ground and then entered the cabin, Romeo style.

Inside, he had huffed and puffed, knocked everything off the kitchen shelves, pulled some red curtains off the front windows, batted a red ornamental jug off the welsh dresser, chewed a red cushion to shreds, and scratched a large red-lettered calendar off the wall. This bear was excited by red.

He did a good job of vandalism, pushing all the chairs and settees out of place. And instead of going out through the same smashed window he came in by, he chose another window, smashed it, pulled it inside the room and then burst out the screen.

Bears are at one and the same time the most secretive and the most stupidly daring of all animals. They have to be secretive to have survived in the east as they have against all the hunters and trappers of the past half century in which the high-power repeating rifle has vastly multiplied the chances against anything surviving. In the same era, the gas engine and the outboard engine have facilitated fast travel in the wilds by ever-mounting armies of hunters and trappers. But the black bear still survives in swamps and remote wilderness regions.

From these, he sorties out now and again to play hooligan with the cabins, shanties and cook houses of his lordly persecutor.

Yodel

A lady in a neighbouring city wrote as follows: "Knowing your interest in birds, I wonder could you help me identify a bird that has recently come into my garden and the adjacent trees. I have been unable to see it so far. But it has a queer song. It yodels."

Now, seldom having heard a bird yodel, I immediately wrote one of my fellow members of the Canadian Audubon Society who lives in that city and asked him to call on the lady and try to identify the bird that yodels.

He wrote back:

"I called on your friend, who is a lady of some consequence with a fine home. The bird that yodels is a Baltimore oriole. The sweet warble and whistle of the bird appears to this lady as a most disagreeable noise. When I told her the oriole's song was much prized by tens of thousands of people lucky enough to have them in their gardens, she insisted that all it did was yodel, yodel, yodel, and that it woke her up early in the morning, yodeling right in her window. Surely, she said to me, we have a right to expect, in cities, that wild birds won't come in from the country where they belong, adding to the disturbance we already have to put up with. She wanted to know if there were not some way she could get rid of the oriole from her premises. Perhaps there was some city department that would look after it for her, as it was a nuisance and got on her nerves."

I have in turn written a letter in reply to the lady's first communication.

"The only suggestion I can offer is that you persuade your neighbours to cut down the trees in your city block, perhaps for two or three blocks around, and remove bushes and other offensive harbouring places from your several gardens. If you are unable to achieve this, there are apartments of some sort to be had down in the warehouse and factory districts of your city where you would be spared the sounds of wild birds."

Curiosity

Curiosity killed many a thing besides cats. The curiosity of wild creatures is boundless. An Indian showed me a trick that has served his nation well for centuries. On my remarking that there did not seem to be much in the way of wild life in the area in which we were fishing, he smiled and offered to take me for a walk.

We left our camp after supper and walked along a ridge for about a mile and a half. The Indian led at an astonishing pace, making no attempt to go as quietly as Indians are supposed to go. He stepped on sticks, noisily brushed branches out of our way, talked to me in a normal tone of voice. We were proceeding as if to an urgent appointment.

When I was out of wind and in need of a rest, we halted and had a smoke.

Then my friend led me back across the ridge we had followed, pursuing exactly the course we had come. Only this time we moved with the greatest stealth. And at a snail's pace, halting for a minute or two at a time to listen and watch. In that mile and a half over which we had come without seeing a single wild creature other than small birds, he showed me a deer, two fox, a skunk, five partridges, half a dozen squirrels and a moving flash which he said was a mink, but which I could not identify for certain.

"When we move in the bush," he explained around the camp fire, "every wild animal hides. It stands still. It sees us first, no matter how silently we move. But the minute we have gone past, every wild creature is excited with curiosity. It comes to sniff our trail. It will follow our trail even some distance, trying to figure out what we are and what we were doing. Their curiosity makes them less careful. We follow our own tracks back. And we surprise them."

This is a useful trick to know. On walks in the woods,

go out carelessly. And come back over the same ground slowly and cautiously. Our curiosity about wild creatures is not one two three to their curiosity about us. And probably for good reasons.

Seagulls

Everybody knows a seagull when he sees one. But in the books there is no such thing. There are nineteen kinds of gulls to be seen in Canada, but none of them has the name of seagull. Seagull is one of those popular words that covers all gulls. It is deceiving, because it prevents us from realizing that there are nineteen different species, many of them sensational in character and habits.

A herring gull has lived in captivity for forty-nine years. It is not unusual for them to live to twenty. This is a familiar gull in most parts of Canada, following steamers on wings with a spread of fifty-six inches – nearly five feet, though you do not realize that when watching them floating through the air. The herring gull has an enormous range over the northern half of the earth, including America, Asia, Africa, the Arctic, Europe.

The great black-backed gull is the largest, with a wingspread of five feet five inches, a fierce old tyrant of a gull that subsists on the eggs and young of other gulls in the colonies where it lives, and will catch and eat any smaller sea birds it can swallow. The black-backed has gone north to the Arctic from most parts of Canada by mid-May. The glaucous gull, sixty inches of wingspread, is another Arctic pirate, which lives practically as a bird of prey.

The nineteen species of gulls include a range of smaller birds down to the little gull, only eleven inches in length from tip of bill to end of tail, and the dove-like Bonaparte's gull, thirteen inches; both of them have a purity, grace and style about them that pleases something deeper than the eye in the beholder.

In the summer months gulls are familiar sights to millions of us in Canada. But hardly a thousand out of a million of us will realize that there are nineteen kinds of them to be identified. They will all be just seagulls.

Job for Crows

An old man of eighty, who keeps his eye on the more significant things in life, tells me that there are ten times more crows in the world today than there were three-quarters of a century ago.

"Crows had no particular job," he says, "when I was a boy. Now, they've got a highly important job. They're the clean-up crew.

"Back in the eighties, every farm had a few scarecrows scattered over the fields. You never see a scarecrow now. In those days, the crows had little scavenging to do. Here and there in the bush or the woodlot, a dead cow or a sheep, a deer or a rabbit. So they turned to the cornfields when their job was tidied up.

"Today we should welcome the sight of the crows thronging north. They're the sanitation squad. Along all our highways and sideroads there is a regular charnel house of dead creatures. We see only a small fraction of the birds and animals that are killed by traffic. Nine-tenths of the victims get off into the underbrush or the fields to die. When the frost goes, they present a problem in public health. The crows attend to it.

"And they attend, all summer long, to the same job, cleaning up the mess man makes in the destruction of animals, birds, reptiles, insects, by the ton, along the thousands of miles of road."

This is a crow of a different colour.

Nipped

A Christmas tree farmer who had planted a couple of acres of seedling Scotch pines was disgusted to find that some animal had got into the plantation and nipped off hundreds of the little eight-inch seedlings. With an oblique stroke, each seedling was cut as if with a sharp knife. The tops were left where they fell, uneaten. It looked like sheer animal malice, and the grower decided to keep watch for the culprit.

It was culprits. And they were rabbits. Who would suspect a rabbit of malice?

But a biologist called in to advise in the matter had an interesting suggestion. A pine forest, when mature, has little or no undercover. Rabbits find poor pickings in pure pine stands. Therefore, rabbits are opposed to pine forests.

"Do you mean to tell me," asked the tree farmer, "that rabbits have enough brains to figure a thing out like that?"

"No," agreed the biologist, "but nature has. And it is possible nature has instilled into the rabbits a hereditary warning mechanism that inspires them blindly to nip off seedling pines whenever they find them in quantity."

The Beaver – Eager Engineer

Due to the pretty general adoption of new regulations protecting the beaver across Canada, many summer cottagers and tourists are becoming as well acquainted with the beaver as they are with the porcupine. In areas in the east where beavers have not been seen in two generations, they are now becoming fairly common. And many a cottager will arrive at his summer home to see about the familiar shores the tell-tale stumps and chips of poplar, alder, willow and other trees. The untidy, heaped-up lodge is not hard to locate, and if no lodge is in evidence, a home dug in an earth bank of pond or stream should be looked for.

Beaver become very accustomed to human company, and can be observed from vantage points not too distant for naked-eye view. The beaver is eager, all right, and he is also capable of mighty ambition. At my summer place, there is a river in which a rocky narrows occurs and a strong current flows. This narrows is just large enough to permit an ordinary pleasure-cruiser to pass through it. On infrequent visits to the cottage in early spring, a fine big old beaver was regularly observed sitting on a rock at the upstream end of the narrows. When boatmen came by, he would slide off the rock and vanish. But whenever seen, he wore a preoccupied and thoughtful expression. He was an engineer with something on his mind.

Then one weekend we noted with astonishment that both shores of the river, upstream from the narrows, were lined with beautifully cut poplar logs about twenty feet long, all trimmed of branches, and at least thirty in number. Each log was laid ready, one end in the water, to be skidded into the stream. We had interrupted the old dam builder at the very moment he had his materials on hand, and was about to build a massive and solid dam across our water highway.

One of the neighbours pinched his logs and towed

them away to cut up for firewood. The old engineer disappeared. But the chances are good that he is sitting quietly somewhere else, nursing his idea. And as soon as quiet once more descends on our neighbourhood at summer's end, he will try it again. If he succeeds, he will make a beautiful mess of a considerable area of cottages and their tidy beaches and wharves.

Goshawk

Watching out over the windy wide bay in front of our summer cottage, I saw a crow coming from the far shore. Instead of being at the usual thirty- or forty-foot height at which crows usually prefer to fly, this one, with that peculiarly ragged flight of the crow, was coming not more than three or four feet off the rough, tumbling water. The gusty wind buffetted it, and several times it appeared barely to miss being doused – a bad thing for a crow.

On it came until it reached the shore, and it dove straight into the pines and scrub.

In a moment, I saw another crow coming the same route, again barely clear of the waves. I watched its laborious progress through the turbulent air a yard or so above the white caps. It too cleared the beach low, and vanished into the scrub.

I was still staring over the water, wondering what strange reason the two crows, members of a particularly meaningful tribe of birds, could have for such a curious performance across a mile of water, when, from a high bluff to the west, there appeared another bird: a goshawk.

Blue-grey, beautiful in pace, quick wing-beats and short sails, it crossed the bay at a respectable height of forty feet.

And the crows in the scrub made never a sound.

Outboard Alarm

You will often hear it said that what has ruined the fishing is the outboard motor. It has allowed sportsmen to penetrate every nook and corner of our waters. As motors became plentiful, Len Hughes of North Bay, known as the Bull Moose of the North, an old-time pioneer of the outfitters' trade, scoffed at the idea that the outboards were ruining the fishing.

On the contrary, in Len's view it has been a boon to the conservation of our game fish. The sportsman in his outboard skiff races to a good fishing spot. He turns off the engine, and the waves of his approach scare every fish within a hundred yards. The clatter, as he prepares his tackle, adds to the turmoil. Casting furiously for a few minutes, maybe ten, he decides there are no fish here. And with a great whoosh and clatter, he starts his outboard and away he goes to another good spot.

The tough days for game fish were those in which the sportsman approached the good spots by canoe or rowboat, cautiously, stealthily, giving the fish no warning.

The outboard engine is the fishes' best friend.

Fish Frivolity

We do not think of a fish as amusing itself. There does not appear to be much humour in fish. They lead cold, purposeful lives, either holding themselves eternally in position in fast rushing water, like trout; or hiding gloomily in the shadow of weed, stump or rock, like bass. If they jump a little gaily on occasion, we have supposed they were catching an insect on the surface or perhaps escaping from a pursuing enemy.

One of my friends witnessed a very curious disturbance that suggests small fish, perhaps young fish, do enjoy a little fun now and then. A chip of wood was floating slowly on the surface of a stream. As it passed near him, in his canoe, he was surprised to see a considerable, very small disturbance around the chip. The water was busily agitated on all sides of it for a foot or more. And when he focussed his attention on it, he was astonished to see a very small minnow spring out of the water, an inch or so from the chip, land on it, bounce about for an instant and then flip off back into the water.

He paddled very cautiously closer to the floating chip. The water was fluttering and dimpling all around it. Small fish were leaping and half leaping in the water. And every few seconds, another would leap successfully onto the chip, take a short ride, and then flip off again.

As the chip was drifting towards shore, where some rocks afforded a little height for a good view, my friend went ashore and took up a position from which he could look down into the water. There was a school of forty to fifty small fish, which he believes were minnows, following the chip, milling eagerly about it within a radius of a foot of it, and jumping onto it for for no apparent reason but from sheer fun.

When the chip grounded, he rescued it and examined it to make sure there were no insects or other things adhering to it to account for the excitement of the minnows.

It may be a case of fish simply amusing themselves.

Skunk

July is the time of year when young inexperienced skunks are finding out about motor cars, to their mutual sorrow. The highways of the country are beginning to be marked with the flattened reeky remains of one of North America's most widely distributed animals. Mighty few animals enjoy a transcontinental range like the skunk's. Even the muskrat, which is known in the wet Louisiana bayous as well as in the arid sloughs of the far prairies, cannot compete with skunks for universality. The red fox, the cottontail hare and others that are imagined to be everywhere have very restricted territories in comparison. The striped skunk, which is the one we know, starts far south of the Mexican border and ranges right up, the full width of the continent, to Great Slave Lake and the tip of James Bay and out across Labrador. The way to get on in this world is to be highly offensive.

But there are four blank spots on this continental skunk map. Canada's tenth province, Newfoundland, is free of skunks. So are Cape Breton, Anticosti and Vancouver Island. The only other blank spots are the northern area of British Columbia and the California peninsula. Any of these unique areas might adopt the skunk as their emblem, surrounded by the golden legend: "We don't have any!"

Two of our characteristic northern animals are the skunk and the porcupine, both of which enjoy a tremendous distribution, probably because they do not have to worry much about interference from any source save man. The porky is luckier than the skunk, since only the wolverine and one or two other daring animals have discovered the trick of killing him. But a skunk is often snatched up by a great horned owl. Hungry wolves and coyotes driven out of their better senses will endure a stench for a meal. Or maybe they like an occasional dash of spice in their food. Lynx, badger and eagle are known to take the occasional skunk.

Trappers collect a fair number annually, though the price of a skunk pelt is hardly worth the trouble. And the reputation of skunk oil as a cure for rheumatism has declined, to the advantage of the skunk tribe. So the motor vehicle stands forth now as the chief enemy of skunks, as it is also one of the chief enemies of man.

The Wasp

A wasp got caught in a spider web and its savage struggles attracted me to watch the proceedings. Here was to be a battle between two members of the animal kingdom, neither of whom is inclined to take a kindly view of his fellow beings.

The wasp's immediate instinct was not fear but fury. He was not badly caught in the web, and his struggles were maniacal as he jerked and twisted to free his legs from the sticky cord, his abdomen bending downward and writhing to present his sting to the enemy he could not see.

The enemy was a small, pudgy spider who behaved like an anxious housewife with a washing on the line, and it beginning to rain. She dashed out of the web and stood unsteadily watching the wasp's corner of the web. Then she would race back to her hiding place up on one corner of the web. Out she would come for another progress report, on the dead run; and after a moment, race anxiously back to her kitchen. According to the rules of spiderdom, the wasp should have become the victim of its own fury, and bounced and jounced in the net violently enough to finally entangle wings as well as legs. But luck was against the rules, and the wasp began to free its entangled legs. The housewife came rushing out three or four times, and you could almost hear her anxious mutterings and "Oh, dears!" and "My, mys!" as the tidy web shook and heaved with the wasp's violence. Just before the wasp freed itself, the spider ran perilously close and tried to throw a new thread of her silk in such a way as to float it across the wasp. But the writhing, threatening sting of the insect unnerved the arachnid, and she backed away and made another hustling run to her far corner.

With abrupt suddenness, the wasp broke loose. My face was about eighteen inches away. When a wasp is worked up, it loses all sense of proportion. It flew straight as me, an innocent bystander, and stung me with complete gusto fair on the chin. The moral is: always watch fights from the two dollar seats.

Ferret, the Pacified Weasel

A ferret is very well thought of. In the hands of a skilled poacher, it can be trained to go down rabbit burrows and chase the game out into the waiting bags or nets of the hunters. So useful has this little animal been over the centuries that it has given a verb to the language. And to ferret out the facts is to go after the facts down all the burrows and hidden places in which facts are likely to conceal themselves, especially in ledgers, blue books and statistical reports. A man who can ferret out the truth is most highly regarded by his fellows.

A weasel, on the other hand, does not work for man. The difference between a weasel and a ferret is actually very slight. They belong to the same family. They might both be called weasels, if you like. But since a weasel is a wild creature, working only for himself, he does not enjoy any reputation at all. He goes down rabbit burrows and not only chases the rabbits out, but catches them and eats them himself, without any permission from any man whatsoever, whether a poacher or a highly respectable gentleman like a broker or a chartered accountant.

To call a man a regular ferret is to compliment him.

Any chartered accountant would be flattered to be called a ferret. But to call a man a weasel is to insult him. If you call a broker a weasel, he is liable to sock you.

Badger

Western wild creatures are moving east. For some years, the coyote has been cautiously venturing eastward off the prairies into Ontario and Quebec.

The new invader from the west is more surprising. He is the badger. While badgers are commonly thought of as prairie dwellers, preying on the ground squirrels and prairie dogs as their lawful meat, they are distributed pretty solidly from the Pacific coast across half the width of the United States. And the Ontario invaders are supposed to have swum across the St. Clair or the Detroit River.

Badgers are members of the weasel family, like the otter, skunk, wolverine and others. They might be described as the lowest slung, most bowlegged of the weasels. They are extremely powerful animals and weigh up to 25 pounds. They are yellowish-grey in colour, with white cheeks and a white line from between the eyes running a short way back over the head and shoulders. Their claws are long and powerful, and a badger can dig like a bulldozer. One of his tricks when sore pressed is to dig in and pull the hole in after him, vanishing.

His normal food is ground squirrels, but it may be Ontario's lush population of groundhogs that attracted him.

The Pine

The resort owner gave us our directions as we pushed off in the outboard skiff from the dock.

"Go down there about a mile," he said, "until you come to the pine. Then turn right, in the channel, and keep on until you come to a round bay. That's where you'll find the fish."

None of us asked which pine. There were a thousand, maybe two thousand pines in and along the shore as we travelled the mile down the bay. But that is the character of pines. I doubt if a higher compliment could be paid a tree than by that resort owner and us three fishermen in the skiff.

For sure enough, about a mile down the bay, we saw ahead of us a particularly spectacular pine on the right-hand shore. It was not a big pine. We had passed, I should think, a hundred much bigger pines in the mile we travelled. But this was a pine of character, as so many pines are when called upon to stand forth in any special fashion.

"The channel to the right," we said, "will be in by that pine."

And it was. We could not see the channel until we were within a couple of hundred yards of it. But when the resort owner said "the pine," there was no mistaking his directions. This, out of thousands, was the pine.

Canada is blessed with a great many trees of character. The elm in the east, the fir in British Columbia, the eternal humble little poplar of the prairies: all have character. But the pines seem to steal the show as tragedians, comedians, character actors. We have nine species of pine in Canada, and one variety of a species. In the east, it is the white pine that made history and is still the ruler of the clan. The red pine occupies much the same territory as the white, starting in Newfoundland and coming as far west as mid-Manitoba. There is a variety of white pine in British Columbia that reaches 250 feet in height and a diameter of eight feet. But it is the jackpine, the least respected of all pines, that does its best for Canada. It starts in Newfoundland and runs straight through the wilderness through all the prairie provinces and right up into the Yukon. In dense stands, the jackpine is a pretty hundrum commercial tree. But out alone on a crag or a sand spit, it is a tree for artists to paint. In my elder years, it is jackpines I am planting, as seedlings, in the places I would like to leave some memory.

Healing Scars

A back country road was converted into a section of a new highway, and the construction crews wrought havoc for miles along its formerly pleasant woodland way. The bush was razed on both sides, bulldozers ripped and reshaped the banks and verges. Rock cuts were blasted. To drive it, after it was paved, was a desolate experience, and those of us who used it felt it would never be the same sylvan passage again.

On our first visit the next season to the summer cabin, we took the familiar route. And although spring had barely laid its hand on the woods as yet, it was simply astonishing what healing of the scars had already been effected by nature all alone. The trees along the way, outside the wrecked right-of-way had begun to reach their branches out into the space vacated by their slaughtered neighbours. Underbrush of half a dozen species of trees and shrubs had sprung up several feet tall. The blasted banks were already carpeted with more than a dozen plants, some of them weeds, many of them native wild plants including such things as the hepatica and fawn lily, as we ought to call the dog tooth violet or adder's tongue. Here and there even evergreen seedlings had nipped hold. Until some engineer with no sense at all save distance in miles and miles per hour sends along the spray truck to poison the roadsides, nature will work in her calm, unruffled fashion to reclaim every foot of the way. And in five years' time, by when the pavement will have begun to decay and the engineers will be busy building new roads elsewhere, you will be as happy as ever, as you drive along it, in the illusion that you have got away from busy little man and all his marvellous developments. The hermit thrush was back, singing, if not nesting, within fifty yards of the already vanishing scars.

Nature appears to be oblivious of man and his developments. Green, yellow, blue are her pet colours. She detests man's grey.

Sentient Plants

It is absurd of us to suppose any longer that plants have no feelings. We members of the animal kingdom have been too smug in our assumption that we alone can suffer, rejoice, tingle, smile, and recoil. This has been because we alone holler, howl, meow, whinny, bawl and bleat, so expressing our feelings. We have even rated the animal kingdom in accordance with its capacity to express its feelings, a bird being far less sensitive than a pig, because it can only sing and twitter, whereas a hog can squeal and grunt in a great variety of tones.

But evidence is rapidly piling up that all living matter is sentient; that is, it feels, having the power of perception by the senses. A plant therefore perceives, and responds to its perceptions in growth, behaviour and survival just as the animal kingdom does. A tree can't cry out when you chop it. But in its fashion it suffers. For it had the senses to respond to the seasons and the forces which made it grow. Dim, inarticulate senses; but feelings, undeniably.

This should be bad news for cows who do not dream that they hurt the grass they crop. But cows do not think of it, any more than we think of the cows when we eat them.

As the old fish wife said about the eels she was skinning alive: "Aw, they're used to it. I've skinned thousands of 'em."

Mink

The scream of a mink is one of the most extraordinary sounds in nature. To my ear, it is sometimes a scream, sometimes a squall. But it is short, sharp and thoroughly blood-chilling.

In winter, when you are walking along a frozen streamside where you are likeliest to see the greatest variety of wild creatures, the scream will bite across the white stillness with the effect of a meteor striking across a black velvet sky. In summer, cottagers on their verandahs are often chilled to a shocked silence when, from the water's edge, that brief, incredible little sound spits.

The mink uses the scream as a warning or a mere ejaculation when it is surprised. When cornered by a trapper, it bares its teeth in a fearful magenta grin and screams. You have the feeling then that it was a blessing that a kindly providence made mink no larger than they are. The cat family produced the lion and the tiger. Thank heaven the weasel family did not evolve a two- or three-hundred-pound mink. The folklore of mankind would have been a far grimmer story than it is.

To give them the will to survive, all creatures have to have enemies. Muskrats are a pet dish of the mink. Mink also love speckled trout, birds, mice, chipmunks and clams. The way the weasel hunts on land, the mink hunts in and near the water. He is a superb swimmer and diver, which accounts for his matchless fur. A hundred years ago, mink fur was just as highly prized as it is today. Samuel Newhouse, inventor of the trap that bears his name and which has caught millions, perhaps billions of animals since it was invented a hundred years ago, writes in his book published in 1865 that he interviewed John

Hutchins, most famous trapper in New Brunswick and Quebec a century ago. Hutchins gave his record of wild animals killed as: "100 moose, 1,000 deer, 10 caribou, 100 bears, 50 wolves, 500 foxes, 100 raccoons, 25 wild cats, 100 lynx, 150 otter, 600 beaver, 400 fisher, mink and marten by the thousand, and muskrats by the ten thousands."

All creatures have their enemy.

Practice

A camping party were at supper near the waterside when they noticed a violent commotion a few yards off shore, and they got to a high rock from which they could peer down into the water. A muskellunge, which is a pretty masterful fish, especially when it weighs thirty-two pounds as this one did, was playing with a walleye, or doré, which later proved to weigh six pounds.

The monster fish would dash in and seize the struggling walleye, give it a few shakes, and then literally toss it away, retiring eight or ten feet to watch. The lesser fish would weave about, trying to recapture its senses. Then the big muskie, backing away a little, would make another violent drive and seize the walleye, shake it mightily, and again give it what seemed to be a powerful toss through the swirling water. After nearly a dozen attacks, the walleye expired, and floated helplessly onto the shore. The muskie idly turned and vanished into the depths. The dead fish was weighed and measured, and its terrible wounds studied.

A good hour later, when an angler with proper tackle came by, he was invited to cast over the spot where the muskie had appeared. He caught it almost immediately, and after a furious battle, it was landed and weighed. Thirty-two pounds. The impression the watchers had, from the whole performance, was that the muskie was merely playing, or possibly practising on the lesser fish. It certainly did not want it to eat.

An associate of the American Ornithologists' Union reports seeing a prairie falcon performing play or practise exercises with a piece of dried cow manure, about the size of a robin. The falcon would fly to a hundred feet, drop the toy and then wheel and dive, to catch it before it had fallen twenty-five feet. After practising a score of times at this, the falcon landed on the ground and practised flinging the toy ahead six feet or so, and then fluttering after it to catch it almost before it touched the earth. Like a golfer practising approach shots, the falcon kept at it until satisfied with its technique, and then flew away.

Among animals, there may be a sense of play, or else an instinct to practise their hunting skill.

Gannet

The gannet is a living projectile. It is a sea bird confined entirely to the vicinity of the deep ocean. It is about three feet in length from tip of bill to end of tail and has a wing spread of about four feet.

Its wings are pointed, its neck is long and very muscular. Its body is compact and almost furred with feathers. From a height of sixty or more feet, it dives at full throttle vertically straight into the sea, a living projectile that penetrates to great depth. It shoots itself at the herring.

Though it enters the water like an arrow, it still flings spray ten feet into the air. Motion pictures of it in action show that, just as it enters the water, it folds its wings not against the body as they are folded when the bird is at rest, but extended back along the body, their pointed length making a sort of sheath, a streamlined sheath. It may not use its wings, the way a cormorant does, to swim under water. But there is reason to suppose that the powerful pointed wing tips may be used as vanes, delicate steering vanes for guiding the course of the projectile as it zips through the water with the momentum of that sixty-foot plunge.

Off the Gaspé coast, on the cliffed islands, and all up the North Atlantic coast, there are colonies of countless thousands of these gannets that survive despite the fact that most of the couples rear only one young a year. They have a curiously angular appearance in flight, yet they impart an impression of great beauty and grace. The Audubon Society and the Canadian government both have fabulous motion pictures of this number one plunger in all nature.

Big Fish

The largest fish are nearly always caught in August. Look over the records of famous fish, and you find generally they met their doom in August. The hale winds of this month may have something to do with it. Or fish may, like humans, feel some presentiment of approaching autumn and decide they ought to stock up.

A big fish is usually a lazy fish. Being close to the reptile family, a fish will often gorge himself and then lie lazy and half-dormant for two or three days. He does not feed continuously like a bird, nor by routine like humans or cows. When the August winds begin to blow and beat the shores with heavy waves, there is wonderful and easy feeding for the big fish. For along the shores, the young fish of many species are buffeted and sent wandering by the waves. Crawfish, frogs, leeches, large insects are loosened from amidst the boulders and crevices and old logs. When a big, lazy fish that has been brooding in the depths all July sees, hears and feels the vibration of the rousing winds on the water, he comes in from his lair to feed up against the coming lean season. Easy pickings. Maybe he is too big to bother with the hors d'oeuvres the waves are jarring loose from the shore. But he finds something to his taste among the smaller fish that have been attracted shorewards by the hors d'oeuvres.

So it is that as he cruises, heavy and choosey, amidst the August fall fair of fish large and small that congregate along the shore, he sees something strange, bright and glittering so boldly darting about amidst the throng. "Oho," says the big fish, shouldering forward, "now what's this here thing!"

He takes hold of it with the only hand with which he is equipped, which is his mouth.

And he finds out.

Tame Fox

A rural friend of mine had a pet fox. Its mother was trapped and crawled into its den, dragging the trap. The farmer dug it out, and in doing so found the cubs. He killed all but one and brought it home, it was so pretty.

In no time at all, the cub was tamed by ordinary farmer kindness, the sort of kindness so skilled in handling colts, calves, lambs. They fed it bread and milk, cereals, egg and scraps, but never raw meat. Whenever they let it have a meal of meat it became immediately spiteful and lawless. Otherwise it was like a cat around the house, romping with the farm dogs whose rougher manners it subverted by its incomparable nimbleness. It throve and grew. It took possession of a shelf in the kitchen as its home and bed. Unlike a caged or penned fox, it had no rank odour. It could be handled and petted; and when the dogs got too rough, it would run and leap into the arms of its human friends.

One day in the autumn it vanished and its masters presumed it had heard the call of the wild. But mention of its loss was made on the local radio. A woman immediately telephoned to say she had heard a lively scratching on her farmhouse door one evening, and had opened the door. A fox leaped inside and bounded on top of the welsh dresser where it curled up, while some hounds that had been chasing the fox raced howling into the farmyard and up to the door.

This was the pet fox, of course; for when my friend called, the fox leaped into his arms with every sign of delight, and was carried home.

But the taste of freedom, with all its adventures, was too much for the tame fox. A few days later it was gone again. This time a week went by before word came that the fox had leaped into the wagon of a farmer ten miles away, again ahead of a pair of hounds. Once more it was brought home. It stayed two days, then vanished for keeps. One of the prices of freedom is peril.

Treed Muskeg

Those of us who love the wilderness for its own sake and not for the fish or the game that still survive in it have reassured ourselves in the past with the thought that in some parts of Canada there are impenetrable or useless or entirely waste areas which will defy the termite-like march of man and survive because of their unprofitableness.

This was a vain hope. If anything in Canada looked secure from man, it was the muskegs which stretch across almost the entire north. Now listen to this, from Ontario:

"A hitherto untapped source of forest revenue has been disclosed by experiments in utilizing types classified as treed muskegs, and considered non-merchantable for pulpwood. In one area in the Fort Frances district, 27,280 trees have been taken off muskeg with a revenue to the Crown of $2,946.

"Permits were issued for the cutting of Christmas trees from treed muskegs or sites incapable of producing trees of merchantable size. An average of 180 trees per acre were removed from stagnant swamps and such lands where trees would never develop into commercial sizes. The majority of them were under two inches diameter. The average age of the trees cut was 110 years."

On commercial Christmas tree plantations down in the farming districts of the provinces, a merchantable tree of better than two inches diameter in the butt can be grown in five to seven years. These little trees of the north have taken more than a century to achieve Christmas tree size. It was in their poverty that their best hope of survival lay, and with it the hope of sentimentalists that an element of untampered-with wilderness, however poverty-stricken, would survive for the comfort of such creatures as might dare to share the earth with man.

It may be said that every tree in Canada has been measured. We are already measuring the midgets.

Nature's Way With Worms

Being conscienceless predators ourselves, we human beings naturally suppose that when worms attack a tree, they are preying upon it. The tree is the victim.

Might it not be that the tree signalled to the worms for help? In the chemical process of a tree there could be some arrangement whereby the tree, anxious for better nourishment around its base, could attract the necessary insects to come and convert the foliage into manure. A tree has two uses for its foliage. With the leaves' chlorophyll, throughout the foliage season, the tree performs the chemical processes which digest the tree's food for its growth, from water, air, sunlight and substances, extracted from the soil. Then, in autumn, the tree drops its dead foliage to serve as a cover on the ground for its roots, to conserve moisture, deflect frost, and in time to rot into compost necessary as organic matter for the tree's further growth. It feeds on itself to that extent.

A handsome white oak near my house, which stands alone, was infested during the summer with a good big species of worm. It did not consume all the leaves, as the tent caterpillar does, but only some. And this worm's droppings, about the size of number seven shot, covered the ground every morning. From a bit of pathway, I could have swept up a cupful from a square yard or so every day. This conversion of green leafage into a type of manure that would be caught in the trash and grass on the ground and penetrate every cranny instead of blowing away, would be of the greatest value to the oak. No other oaks in the vicinity were infected with the worm: only this one standing apart and probably in need of more nourishment. What makes me suspect that this infestation was for the tree's good was an experience some years ago when the whole forest in our vicinity was attacked by a plague of the forest tent caterpillar. Early in June, they simply denuded the trees, all second growth, of every vestige of leaves, so that the bush looked as it does in November. Yet, within two weeks, the whole bush had sprouted new foliage, looking none the worse for its experience.

And the following summer, I never saw such luxuriance as this bush exhibited. It seemed to me that the tons of green foliage that had been converted into manure by the worms had done the second growth bush a wonderful good. A need had been served.

Trees in August

August brings the time of the year when many of the things which beguile us in wild nature are ended. Most of the wild flowers are over; the birds are through nesting and many have ceased their song.

Only modest and spinstery flowers remain in the woods, the plantains, the ladies' tresses. In the fields, the weeds are coming into bloom. What birds are to be seen are the dull and frowsy young, hard to identify, not very amusing to the eye.

But the trees remain. August is the month of trees. They are coming into their mature glory and greatest beauty under the hale winds of August. In spring, the trees give us our first full sense of the annual resurrection when they commence that breaking, as if by stealth, into leaf – not green, but yellow, really. By stealth, in those few days, the trees cover the dirty earth with a haze of yellow, yellowish-green, then green. We might fall in love with trees at that moment of the year, if we were not distracted by simultaneous miracles on the ground: the first flowers. We are all for the little things, in spring.

But in August the mood changes. The trees are strong, their leaves are tough and leathery after three months of increasing sun and wind. A big elm, a maple or an oak, a pine, with its black layers and ledges, tossing and fighting the breeze, are subjects for full and lengthy reflection.

Most of them are older than we are; and will live long after we are gone. Some of them stand where they stood a century ago. Some of them will still stand where we see them, fifty years after nobody in the whole world remembers us at all. This is the mood of August in which to gaze at trees.

Starling, the Brash Usurper

From one point of view, the two greatest disasters that have befallen North America were the introduction of the white man and the introduction of the starling. Both are making a mess of an otherwise beautiful continent. Both seem to think that it is their right to move in and take over the whole business.

There are of course things to be said for both the white man and the starling. I will leave the defence of the white man to those better qualified and less prejudiced, and submit a few remarks on behalf of the absurd bandy-legged, short-tailed, large-beaked bird which in many parts of the country has come to be regarded as a very serious nuisance.

In the first place, unlike the white man, the starling did not come here of his own free will. He was introduced.

Mr. Eugene Scheifflin liberated eighty birds in Central Park, New York City, on March 6, 1890. Forty more were liberated on April 25, 1891. They were imported from Europe and were liberated in the hope that they would control a pest of caterpillars in Central Park. From those 120 birds are descended all the millions upon millions of starlings that now inhabit the American continent from Mexico to the tree line.

If you like symphony music, you cannot help but be spell-bound by the performance of thousands of starlings, like smoke clouds in the sunset sky wheeling and gyrating in almost majestic formations, the thousands of them moving in perfect harmony as if one mind alone governed their incredible dance of worship. If you are bewitched by courage, then you have only to hear a solitary starling in the dead of February, sitting in the bitter sunlight in a tree, going through his repertoire of mimicry, his absurd beak lifted, his neck feathers shaggy, his wings pendant, as he recollects phrase by phrase the songs you have all but forgotten, the songs of robin, peewee, thrush and meadowlark. He remembers them all. He beckons to spring.

But like the white man, to quote a government book, "his unsightly nesting habits, competition with others and his proneness to form immense roosts," counterbalance his virtues.

A neighbour of mine hates starlings. When a delegation of some dozens of them took up his neighbourhood as their preferred residential district he borrowed his nephew's air gun in a sudden ecstasy of dislike for the birds, and somewhat to his own dismay, succeeded in killing one of them out of a busy bevy on his lawn. He was grateful to think nobody saw him. But somebody did: another neighbour of ours who is a government biologist. He happened to be glancing out his bathroom window and saw the fell deed.

He came out and congratulated his neighbour on his marksmanship. The man was humiliated.

"I hate the things!" he confessed in shame.

On the back fence, the biologist laid the starling out and calmly proceeded to dissect it.

"I just want to show you something," he explained to his slightly nauseated neighbour.

From the bird's gullet and in the first area of the bird's alimentary system before digestion had damaged the contents, he ticked out with his knife point five cutworms.

"Recognize those?" he asked,

"Cutworms," agreed the unhappy marksman.

"How long would it take you," enquired the biologist, "to locate five cutworms in your garden?"

My neighbour, whose garden is his pride, is in the process of reorganizing his attitude towards starlings, not an easy thing to do. He tells me it is like trying to get rid of race prejudice.

Large districts of the farming territory in which the bluebird was a familiar and most welcome resident from early spring until summer's end now never see the bluebird at all, or swiftly passing through as a migrant in search of possible nesting sites. The starling has ruthlessly taken possession of every hollow post, tree and man-provided nesting box. The red-headed woodpecker used to be one of the most striking members of the eastern and central Canadian bird community. Now it is a rarity.

The latest bad news, reported by those who erect nesting boxes for the wood duck in the thirty-year struggle to re-establish this most beautiful wildfowl which was nearing extinction in 1920, is that starlings are seizing and holding even these large nesting sites, and several colonies of wood duck have already been eliminated.

The screech owl is a little bird not much bigger than a baked Idaho potato, but with a very pompous air about him. For some years past, one of my sporting friends who lives on a fairly spacious estate in a small town has been putting up bird houses of many shapes and sizes in the hope of encouraging our native birds to live round his property. It has been largely a vain endeavour, because the starlings have swarmed in and usurped every possible nook and cranny, every bird house and shelter, and have taken it for granted that all the feeding stations set up were provided entirely for their benefit.

Some time ago, despairing of native birds, he set up some larger-holed houses in the trees to accommodate the black squirrels that were fairly numerous in his grounds.

Early in the spring, from an upstairs window, he happened to notice that one of the squirrel houses had some sort of obstruction in its entrance. With his field glasses, he discerned that the obstruction was the proud, round, haughty face of a little screech owl. Eyes closed, ear tufts erect, beak tucked into chin, the little owl's face exactly fitted the entrance hole of the nesting box.

As my friend watched, he saw starlings flutter hurriedly nigh, take one look, and then flee to pastures new.

So far not one starling has returned to the grounds formerly brawling with twenty or thirty of them. My friend is thinking of going into the business of raising screech owls for sale to a starling-weary continent.

Poverty

There are thousands and thousands of people right within sight of us who are so poor that they don't know the names of more than four or five birds. They know a gull, a crow, robin, sparrow and, probably, starling. There are people by the countless thousands all around us who are so poverty-stricken they can't identify one tree, calling a spruce a pine, or a maple an "I-think-it-is-a-maple."

Heaven forbid that they should ever have to leave the highway and the ignorance of their cars to go walking in the woods. But if they did, not only could they not name one of the forty wildflowers they tread upon, they would not even see them.

What I am getting at is that poverty comes in more shapes and sizes than money and property. The slum dweller is no more poverty-stricken, in some respects, than the stock broker or the lady in mink. The slum dwellers have known for some centuries that they are poor. It will likely be another few centuries before the rich can perceive that they are poor too.

Shrew

Millions of Canadians live from childhood to old age without ever having seen either a mole or a shrew. Yet there is not a park or a cemetery, a suburban field or pasture in the big cities and towns where these creatures are not to be found if sought. And in the farm country, where both moles and shrews are plentiful, any number of country dwellers get through life without ever having seen either. We just are not curious. If we see something small and grey flicker in the grass, we assume it was a mouse. So what?

Well, take the shrew, for example. It weighs from a fifth to four-fifths of an ounce. It is smaller than a mouse. It is the most tragically fitful, high-strung, nerve-jerked, restless creature in all nature. It dies of old age at fifteen months.

Yet a lion, a wild cat or a wolf is a softie compared with it. It is one of nature's most savage and voracious creatures. It is almost blind, as moles are, and lives most of its brief life out of sight, if not underground.

This tiny, savage beast secretes poison in its saliva which paralyzes and suffocates its victims, such as mice. The poison is akin to that of rattlesnakes.

It is found all over Canada, from the Arctic to the border. In millions, or possibly, billions. But back to, and including Jacques Cartier, few of us have ever seen one.

Otter

An otter can make a greater variety of sound than most animals. It can quack like a duck and bark like a terrier. It can whistle, spit, yelp and growl. And it can emit the most derisive raspberry you ever heard.

If you take an otter unawares, it instantly dives with a startling splash. But almost immediately it reappears, its seal-like whiskers sticking out in angry consternation, while it spits and makes raspberries at you.

The otter is a spirited animal. Everybody knows about otter slides, where they engage in sliding downhill with all the gusto of skiers. But their way of life itself is playful and full of a sort of hilarity. They weigh from fifteen to twenty-five pounds, are often over three feet long from nose to tip of tail. A big, lithe, sleek animal, it can swim like a seal and make a pleasure out of chasing the fish it feeds on. An otter fishing is like a boy playing water polo, full of splashes, whoops and fun. It is said to kill far more fish than it needs, from sheer devilment, leaving the carcasses on the shore with a bite or two out of the back. Who knows but that the otter, a fine, hearty creature, has a soft spot for the bald eagles, crows, fox, bear and other brothers who also love fish but can't chase them? In our quaint human way, we like to reserve the notion of charity to ourselves alone. However, the more you look at nature with your hands in your pockets, the more you wonder.

In summer, when the three or four young are old enough to come forth from the den with its submarine entrance, the whole family goes travelling far and wide, from stream to swamp to lake, sliding, rollicking all the way. An otter may be charitable, but it likes to lose its young in a country far from its own favourite territory.

Wanderers

When you consider how easy it is for a bird to travel, it is a wonder they stay so strictly within their appointed regions. For instance, there are a great many western birds that are never seen in the east, and just as many in the east never seen in the west. Yet scores of species of them winter in Central and South America, and come funnelling up into North America in the spring. It is a wonder so few of them do any wandering from their respective territories.

Yet wander some of them do. And a great deal wider and farther than just around the American continent. Roger Tory Peterson's book, *Field Guide to the Birds of Britain and Europe,* lists a large number of our Canadian birds that have strayed, probably from the far north at the start of the fall migration, not only as far as England, but to Germany, Denmark, Italy. The blue-winged teal, baldpate, bufflehead and hooded merganser have been collected in Britain, France, Holland. Our little sparrow hawk has been picked up in Denmark. The kildeer has been found in Britain and the common little spotted sandpiper in Ireland, Britain, Belgium and Germany.

Our nighthawk, catbird, hermit, olive-backed and grey-cheeked thrushes have all been reported in Europe, and our junco was found in Italy! Of our warblers, the black and white reached the Shetlands, and the Parula and black-throated green got to Iceland. The red-eyed vireo, our preacher bird, turned up in Ireland. And the white-throated sparrow, singing Canada, Canada, has been collected in England.

The experts believe that these "accidentals," as they are called, these far wanderers, are victims of storms that forced them in the wrong direction when the season demanded that they begin their migration. Or it might be these strayed birds had lost, or been born without, those mysterious governors which guide birds on their path up and down the thousands of miles of their routine migrations.

Of them all, I think that tinkling junco that found itself in Italy, or that white throat in England are the two that might have given me heart failure if, say in the middle of the late war, I had heard and seen them on a tree, exiles too.

AUTUMN

Song Sparrow

The time comes, in September, to say good-bye to the migrating birds. They are on their way. Millions of them, Canadian born, are already gone, exploring field by field, county by county, state by state, the republic to the south of us. Many of them will not stop until they reach Central America by easy stages.

Choose your own to bid good-bye. My choice is the song sparrow. To bid him good-bye is perhaps a little silly, because he often does not leave us at all. He merely goes to some bushy river valley nearby and spends the winter there, quietly awaiting spring. If he migrates, it is probably only a little way south. He is none of your world wanderers, like some of the tiny brilliant warblers that journey to Central and South America or the Indies for their winter residence.

But the song sparrow is an essentially Canadian bird. Varieties of him are found in every acre of Canada, you might say, from Nova Scotia to Vancouver Island. He is one of the very first to sing in spring. When the first robin is on the city house top, when the first meadow lark is calling across the bare March farms, the song sparrow is at his music on fence post, bush top, apple tree or telegraph wire, ringing in the new.

When the meadow lark has ceased his calling and the robins are silent save for their melancholy alarm note at the sight of a summer cat, the song sparrow is still busy on his bush or fence post, tinkling at his music while the wheat ripens, the corn grows stately and the smoke of burning straw hazes the air. In North Africa, in late October 1944, I once saw and heard scores of skylarks going through their ecstatic spring routine of mounting into the sky and fluttering there, filling the sky with song. And in a more recent October, in a field bright with pumpkins, I heard a song sparrow pretending it was April.

We make quite a fuss welcoming them in spring. We let them go in September with scarcely a wave for goodbye.

Sanctuary

The idea of creating sanctuaries for moose, ducks, deer, beaver and other wild creatures that are fast vanishing is spreading rapidly all over America. There have been national and provincial parks for many years that have served as sanctuaries. But now the demand for sanctuaries is increasing. On the Great Lakes, sanctuary areas have been set aside for the black bass. Nobody can fish in those areas.

There is one little creature, however, that is being overlooked. How about some sanctuaries for human beings?

There is an increasing type or sub-species of humanity that desires with all its heart and soul to live, for part of the year anyway, amid the natural, unspoiled earth the way God made it. We have to confess that the majority of mankind at this moment do not belong to this subspecies. Start a phonograph, drop a coin in a juke box, paint a shack bright red and dispense pop – and you will attract 500 people for one who will come when a loon calls on a lonely lake.

According to the rules, therefore, the earth must properly be handed over to these juke boxers and pop wallopers. They are in the majority. Therefore highways must be driven into the last wilderness for them. They fear solitude. They hate silence. Their mission is to carry into the frightening still places of the earth the healing rumpus of cities.

Set up a moose sanctuary: and immediately roads, service stations and dance halls must be instituted in it so that nature lovers may come by car, by outboard motor (of ever-increasing power) and by airplane to see the moose.

There is, however, a dream sanctuary which someday may come, for men and beasts alike: where all engines are prohibited, and man can enter only by foot or canoe, with his food on his back and his canoe under him or on his head. That alone is sanctuary.

Porky

The sportsmen with their dogs are taking the field after game birds and game animals. And one thought is uppermost in the minds of the men in these teams: "I hope we don't meet a porky!"

There must be something peculiarly attractive in the scent of a porcupine to cause a dog to rush to his painful and ludicrous doom the way he does. Well-trained dogs that are expensively taught never to pay the slightest attention to a hare or a fox or any other creature save the game birds their masters seek, will, without an instant's hesitation, dash in on a porcupine. The result is always the same. The dopey, slow-going porky, noting the attacker, simply wheels or pivots on his powerful hind legs and meets the onslaught with a swipe of his powerful, muscular tail radiant with thick, sharp barbed quills. The boldest of dogs, unafraid of anything its size or bigger including wildcats, receiving this violent slap on the snout, lets out one astounded yawp and forgets all about its quarry of an instant before in the agonized pawing, licking, slurping necessitated by the pussful of stinging quills it has received.

Some dogs never seem to learn. They will attack porcupines time after time, with invariable consequences. And the hardest part of the experience is not the reception of the quills; it is the removal of them, by sweating, fighting, cussing masters armed with pliers. When the boss takes hold of his dog impaled by porcupine quills, all friendship ceases. The dog acts like a lunatic, and the master acts like a tough. It is no time for sentiment, for the quills dig in, every passing minute. The best thing to do is wrap the dog in a blanket, pinion him helpless and frothing, and then yank, yank, yank.

Meantime, the porky, artless cause of all this tumult, goes stodgily about his affair. He may hurry a little. But not for more than about ten yards. Then he forgets the incident entirely. The porcupine is a very old and respectable family. He is unchanged today, in the least degree, from the porcupines found in fossil remains millions of years old.

It is hard to believe the ruin the porcupine can wreak. A late fall wind blew a dead branch off a tree which smashed a window in a neighbour's cottage. A whole community of porkies moved in for the winter. They chewed kitchen table legs to mere spindles, gnawed axe and other tool handles to shreds, nibbled shelves, gnashed benches and chairs. The place was a shambles when the family arrived in the spring.

Man is probably the only really efficient enemy of the porky. Give man an economic basis for his enmity, and heaven help the porky.

Disoriented

The steadily increasing sport of fishing reveals one of the most amusing characteristics of human nature. And here is my favourite illustration of it. About twenty-five years ago, a great, big senile old muskellunge, the king of fresh-water game fish, weighing about forty-five pounds, got lost. His normal haunt or holt, which I knew well, was in the shadow of a reef lying at the foot of a rocky cliff. He had been there as long as I, or the Indians who showed me where he was, could recall. It was his home; and when he (or she) went away in the early spring to spawn in some shallow marshy bay of the lake, he always returned to his holt for the whole summer, and scorned all attempts on the part of sportsmen to take him.

But in his old age, though in physical prime, he must have chased a four-pound bass for lunch one day, and lost his bearings. Anyway, he wandered miles from his holt, and was one day caught by a trolled minnow in a channel where no self-respecting muskie would ever dwell. He put up a magnificent battle despite his senile condition, but lost. And was exposed to the vulgar public gaze of the entire lake.

And for twenty-five years, that channel has been a famous muskie ground for hundreds upon hundreds of sportsmen who have combed it season after season, full of high hopes. Not another muskie has ever been caught where the old, lost muskie, miles from home, got hooked.

Caspian Terns

The gull, the herring gull in particular, has for ages been the wonder of man for the beauty and ease of its flight. Poets, artists have watched for enraptured hours across the centuries and tried to capture it in words, paint, bronze. Early experimenters with aircraft mention the gull as one of the great inspirations of their dreams and speculations.

But in the eastern half of Canada, from the tip of the Great Lakes to Newfoundland, there is a bird that possesses something the gulls want. It is the Caspian tern, a bird as large as a herring gull and easily mistaken for it by those who give such things the casual glance. Anything more than a casual glance will note the bright red bill of the big tern, its black cap and, of course, the entirely different design and symmetry of the bird in flight.

A pair of Caspian terns came fishing during a Great Lakes storm off a rocky point where schools of minnows or small fish were being buffetted suitably to the terns' needs. The storm was a gale of possibly forty miles an hour, with violent and erratic gusts half a mile wide whipping and lashing the already mountainous waters. The two beautiful terns rode the storm with a smooth and effortless grace that could be likened to nothing else than skating. Figure skating. Up the gale they would beat their way, almost lazily, their red beaks pointing vertically downward as they scanned the riven waters, fifty feet below them. Every thirty of forty yards, they would lift and dive – suddenly, unerringly, you could tell by the short shake they gave themselves immediately after the dive, as they swallowed their catch.

Into the picture came three herring gulls. Down the gale they came sliding, and seeing the terns, tried to horn in on the feast. The instant they turned, they were like clowns, ungainly, tumbled, ruffled, teetering. In thirty seconds, the gale had blown them away to some sheltered gull haven.

The Fawn

A backwoods friend of mine found a tiny fawn, about two weeks old, staggering along a bush road. Now, while this man will shoot an old and faithful dog the minute it becomes a nuisance and will peel the hide off a beaver or mink without any qualms, he is at heart, like most backwoodsmen, a very compassionate man. He went a distance and watched a full hour to be sure the doe did not find the fawn; and then, assuming it to be an orphan or a twin that had strayed, he collected the fawn, brought it out to his cabin, where, that afternoon, a passing bush pilot picked it up and took it to the summer home of one of the best deer hunters I know among city men.

The poor little starved creature – it had obviously been lost more than the day – was fed condensed unsweetened milk and pablum. It survived. It began to grow. In two weeks, it doubled in size. It slept on a mat behind the kitchen stove, roved the woods around the cottage and came when called. It developed a tremendous taste for pretzels, and would hurl itself at the screen door of the kitchen until admitted and then would mince to the pretzel drawer and wave its ears in a most commanding gesture for so ravishing and ethereal a creature.

My deer hunting friend had seen fawns before, but never played host to one. No matter how many close-up pictures or paintings of fawns you may have seen, you are utterly unprepared for the real thing.

There is an absolutely breath-taking beauty about them, the appreciation of which grows with every hour. Their angularity opposed to their elfin grace touches something deep in your nature. My deer hunting friend was in one of the major quandaries of his life. The game warden told him to leave it behind, free, in September. But meanwhile the host to the fawn quietly removed two sets of deer antlers that decorate his cottage living room. "The fawn," he explained, "will be old enough to notice things in a few days. . . ."

Wild Colour

"We have hardly a weed we can call our own," said the famous John Burroughs. Most of the colour along the roadsides and fields in the conclusive September days is provided by weeds. And nearly all the weeds came here from Europe. Shepherd's-purse, campion, corn cockle, Queen Anne's lace, moth mullein, butter and eggs, sow thistle, the blue weed or viper's bugloss – how's that for a name? – and the teasel the arty ladies love to gild for a dried bouquet; all, all from Europe.

Some of them came two to three centuries ago, like the

common white daisy, in the seeds and fodder the first pioneers brought with them. In seed, in the mud tracked off ships by the immigrants or the cattle, the several hundreds of foreign plants that now call themselves Canadian or American got here and went on the loose.

But we have plenty of native autumn colour besides the native trees. In eastern Canada alone there are some fifty species of goldenrod. And of the 275 species of wild aster the botanists have named in the world, most of them are in America, the majority of them in some part of Canada. And all of these asters and goldenrod choose September to be at their best, as if, being Canadian, they had a job to do with their colour in harmony with the trees.

After the vulgarly simple design of most cultivated flowers, it takes an appreciator of fine things a little while to get the hang of these wild beauties. You must look a little closer at a plume of goldenrod or a spray of wild asters than you would at a rose or a tulip. But if you might be so daft as to carry a pocket magnifier with you when you go strolling in September, there are elements of design and adventures in pattern to be found in these last, lean wild flourishes of nature that can positively humble you.

Horse Chestnut

A couple of blocks north of us is a horse chestnut tree that is a matter of local pride. In spring, when it is lit with a thousand candles of bloom, we neighbours all take a stroll to see it in the evening sunlight. Now in autumn, it turns a barely credible virgin gold, so shimmering in beauty when the last sunlight touches it, that it makes your heart stand still an instant. One day in late October or early November, it will perform a ceremony of disrobing that has to be seen to be believed. Without warning, and at no foreseeable time or temperature, it will suddenly start to shed its golden leaves. Not a dozen leaves will have fallen in the weeks before, when all the elms and maples of the neighbourhood were shedding day by day.

Then, as if in response to some inner decision, the chestnut will simply let go. It is magical. The leaves begin to fall. First a few, then dozens, then hundreds. They stream down, making great marks on the ground. In four hours, the beautiful tree is bare and stark. It is as if a garment, a golden mantle, had been loosened and dropped at its feet.

Hawk Migration

On the fine northerly September winds, no more beautiful sight in nature is to be found than the mass migration of the hawks. Hawks of many species seem to join up for the movement south, the great buteos wheeling on motionless wings, often at great height and in spectacular numbers, with an air of glorious idleness, but being carried on the lofty gales hundreds of miles in a day.

The smaller hawks, the accipiters as well as occasional falcons, are members of this September-end migration. You can spot them, amidst the wheelers and soarers, taking quick beats of wing to recover their balance on some vast rising and moving column of air that is bearing them all up and away to distant lands. With a couple of mice in his craw and half a dozen grasshoppers for dessert, a hawk can make hundreds of miles riding these sun-bright September northers, where, if he were one of the humbler birds that has to march every foot of the thousand-mile journey on his two frail wings, he would have to stop for fuel a dozen times in the distance.

All across America, there are certain air highways the hawks follow on these soaring trips. Mountains, cliffs and lakes are the ground features which create the mighty updrafts or funnels the hawks love to float on. And a fine, sunny day with a fresh northerly wind is the combination that makes for a good fill of hawks for your field glasses.

Bird watchers count these hawk days as red-letter days in their year. It is the most comfortable and inspiring of all the experiences of bird watching. You drive to some high, rolling country, preferably with a body of water to disturb the wind, and there you lie down on a hilltop with your field glasses and start searching the sky. At first, you may not notice more than one or two. But if you hit the day and the place right, suddenly the sky seems full of floating beauty. And all going away. . . .

Venison

As we drove through the beautiful Swan River valley in northern Saskatchewan, a doe and her last year's fawn bounded out onto the highway, dodged gaily ahead of us, white flags waving, and when we slackened speed and stopped the car, they stopped too. They turned and stood looking at us, alert, twitching.

"How anybody," said our driver, who is a geographer and a most dedicated conservationist, "could shoot one of those lovely creatures, I will never know!"

The other two passengers with me echoed the sentiment with murmurs, for it was a bewitching sight: the glorious valley, the ancient spruce hills, mile on mile, and the two deer standing not fifty feet away, gazing moon-eyed at us.

"Beautiful, beautiful creatures," sighed the geographer. "Yet, the hunters swarm through here . . ."

We drove on. And the following noon we were at Duck Lake, scene of the Riel Rebellion massacres, visiting the memorial cairns and the site of the battle. We were guests at lunch of the leading citizens of Duck Lake, all of them ardent sportsmen in that land teeming with wildfowl and upland game birds. And our hosts informed us heartily that in our honour they had got out a haunch of venison and the steaks were to be cooked in a dry pan after the old French-Canadian fashion. . . .

There were platters of chicken, large bowls of salad containing hard-cooked eggs and potato and several other items in the spread that would have sustained life in a geographer.

But no! He, like the rest of us, reached for the great big platter with the venison steaks.

And how many do you think he ate?

Three.

As he slowly chewed the last lovely bite of the third and last steak, he became conscious of my steady gaze upon him, from the far end of the table.

In order to eat, you have to take life, whether a dove or a potato, a lamb or a peach.

Kildeer

Early in the spring, I turned my car onto a gravel side road, and my eye was caught by a tiny creature running with remarkable speed and total irresponsibility in the middle of the road in front of my wheels.

It was a baby kildeer plover, a comical and endearing miniature of its parent. Then yards farther on the road, the mother kildeer was flopping about with piteous calls in the famous broken-wing act of birddom. I waited the several seconds until the tiny creature, surely no more than a couple of days out of the egg (for they are operational from the moment of hatching), ran onto the shoulder and vanished in the grass.

Now, in October, I wonder where that tiny morsel is. For it will be a full-grown kildeer long since, and it may be in Bermuda, Brazil, Peru. The kildeers winter all the way from southern Illinois to South America.

When we think of life, not just human life, but life as a whole, the entire immense mystery, there are instances in the plant world as well as the animate world that stagger us with the furious force of life. The birds, from egg to nestling, to fledgling, to full-grown adult in a matter of weeks, offer us the most spectacular demonstration of life's boundless energy.

Canada Goose – A Nobler Symbol

There is a fairly large body of opinion holding the view that the Canada goose and not the beaver should be Canada's national symbol. The beaver has much in his favour. He is industrious, ingenious, patient and only death can put an end to his constructive activities. But he is a homely, buck-toothed character. He is virtuous. But so is a deacon with a buck-saw.

The Canada goose has all the qualities of nobility. He is purely and almost wholly Canadian. His home is here and when he goes south, he goes only as far as he has to. Vast numbers of Canadas winter in Nova Scotia and at other points barely over the border. Before the ice goes out, in February even, they are homing again.

They are bold, brave, cunning, shrewd. They have an extraordinary community spirit. They are monogamous, choose one mate, and if it dies, they mate no more. The gander will fight like fury to protect his mate and her nest. They are beautiful. They are mighty. They have a wingspread of five to six and a half feet.

They are familiar from the Atlantic to the Pacific and to the Arctic Circle.

And any night in late October you may hear them passing over. They call the sound they make honking; but to my ear, it is barking. And they sound like a pack of heavenly hounds streaming across the sky. Sometimes you are fortunate enough to see them in daylight. Then you hear them long before you see them. When they come into view, high and far, they are very often in a V formation, but sometimes in a long single oblique line, and occasionally in Indian file.

As their slow, powerful wing-beat is not in unison, the line seems to undulate and waver with an eerie, mystical motion in keeping with the wild, weird music that starts and fades, rises and falls as they sail past.

The speed he goes at, effortless, full of conversation, is forty-five miles an hour, gauged by theodolite from the ground.

Most Canadians know who Jack Miner was and what he did with his wild goose sanctuary at Kingsville, Ontario. He trapped and banded the geese, which trusted him, despite such familiarities. Many wild geese bearing the metal leg bands, each with a chapter and verse of the Bible indicated on it to identify the date and place of banding, were taken by hunters.

In southern Michigan, a wild Canada goose was found dead from gunshot, and it carried two leg bands. The first read "32 F, Mark 5:36," and the other, "AS, 44, Mark 5:36."

The finders of the goose reported to Manly Miner, the son of old Jack, and Manly reported that the bird was first banded at Kingsville in 1932, and again, at Kingsville, in April 1944. This means that the grand old gander had lived at least 23 years, and had travelled up and down the continent, spring and fall, possibly from the Arctic Circle down to the Gulf of Mexico, twenty-three times before being shot. And stopped off at Kingsville, from time to time, in passing, to pick up a little scripture.

Pintail ducks are not nearly as long-lived as geese, but are tremendous travellers. One banded in Labrador in 1951 was shot in England, in 1952; another banded in California was shot in Cook Island, New Zealand, three months later; a pintail drake banded in Maui Island, Hawaii, fell to the gun of a hunter a year later, in Edmonton, Alberta.

By the way, the verse that old gander carried up and down the world, from St. Mark, reads:

"As soon as Jesus heard the word that was spoken, he saith unto the ruler of the synagogue, Be not afraid, only believe."

Man the Provider

October is the time of year when a strange uneasiness enters the mind, bones, and tissues of men who live in city and town. They are restless, impatient, with the need to stand and stare into space. They heave heavy sighs.

Country men don't suffer from these vapours. The farmers are weary with the last of the harvest, they are full of the contentment with the long job done. It has been a good crop or a bad. It does not matter. They have performed their function. They have provided.

The trouble with city and town men in October is that they are the descendants of countless generations of men before them who, until quite lately – a mere fringe of time compared to the vast expanse of the past – had ingrained into their deepest consciousness the desperate need to provide in autumn. The city man, whose providing is done week by week, all year through, doesn't know what is the matter with him when these ancient stirrings start to rankle within him.

That is why going hunting affords a mysterious peace of mind to those whose subconscious instinct leads them out into the wilds. For that is where all mankind went across ten thousand, a hundred thousand years.

Those who didn't, didn't leave any descendants.

Clouds of Gunners

In the vigorous fall season some hundreds of thousands of Canadians – according to the provincial issuers of hunting licences – are abroad in the land with shotgun and rifle. Being partial to the chase myself, I wish I could say something pleasant in their favour.

The late Jimmie Frise's mother used to say: "A man with a gun on his shoulder is generally no good." But Jim bore a gun on his shoulder every chance he had. You come by the blood lust *nolens volens,* as the lawyers say. You can't help it. My great-grandmother Duncan is to blame in my case. In a cow's breakfast hat of her own design and manufacture, she would sit all day in a punt, fishing; or in her husband's boots, follow the hound all day after a fox or deer, while the chores waited and her children wailed. But how she could cook a duck! Black duck preferred.

In her day, at this season of the year, the sky was clouded with game birds, ducks, geese, shore birds, pigeons. Food streamed through the sky. With their old percussion cap muzzle loaders, they had not far to go from their farmhouse doors to fill not a frying pan but a great cauldron with game stew enough to keep their sons and daughters equal to the axe and the yoke and the mattock. The mattock was the semi-hoe, with which the ladies, in those days, loosened the new-cleared earth to receive a potato.

There are no clouds of game in the sky today. There is not even a mist. The shore birds that used to darken the sky, with the pigeons, are all protected as rarities. The pigeon is extinct. Wealthy men now travel to the subarctics, by plane, to shoot at the geese our great-grandmothers never lifted their heads to look at, as they marched south, laughing.

The millions of men and women in Canada and the United States who are exercising their licences to shoot, are not shooting for food. They are shooting for fun in a desperate effort to escape from a world in which food comes from cans.

Psh

The art of calling wild creatures is as old as the hills among hunters, but the field naturalists are just starting to take it up in a popular way. Our Canadian Indians were always regarded as great hunters, not because they were wonderful shots with firearms, but because they could get close to their game very often by the device of calling the game to them. The birch bark horn, really a megaphone, through which the hunter weakly whines and grunts like a cow moose, was one of the Indian ingenuities.

But I have seen Indians, by means of diminutive squeaks, hisses, chirps and other barely audible sounds, call weasels and mink out of rock piles, raccoons from their tree dens, beaver out of their houses, and a deer out of a dense alder thicket.

In the west, experts can call the mule deer with a reed-like instrument. The modern duck hunter would not dream of shooting from a blind without a duck call to attract the attention of the passing birds to his decoys. With a crow call, you can pull crows from miles around to their doom.

There have always been field naturalists who could call certain birds by whistled imitations of their calls. Anybody can call chickadees, the white-throated sparrow, the crested fly catcher and the whip-por-will by such imitations. But there has crossed over the border from the New England states a trick for enticing all sorts of birds out of hiding. Field naturalists have used it with great success, paricularly in the nesting season; but it will work fairly well even with the winter birds. It is so simple as to be hardly credible until you see it work. You simply hiss, in a series of short, sharp hisses, "psh, psh, psh," through your teeth. What the birds think it is nobody can say: a snake, an alarm note of other birds, a weasel? At all events, if you utter this small hiss outside brushy cover in which birds are likely to be hiding, they will come out immediately, and usually at close range.

Spruce Gum

Hunters in northern Ontario brought a large black bear they had shot to the local lands and forests officers. It was in fine, healthy condition, but two big black scabs on it suggested that the carcass might be of interest to the biologists. One scab was on the fore-shoulder, the other on the opposite flank.

Biologists being experts on postmortem, an autopsy was held. It appeared the bear had been shot down some time previously, possibly a couple of years. The high-power bullet had perforated the animal without hitting a vital organ. The muscles and tissues had been damaged. On careful examination, the scabs proved to be successive applications of spruce gum, one on top of the other, until the damaged areas had been thoroughly covered, and the gum had hardened to a protective shield.

In the James Bay area some years ago, an Indian suffered a badly broken forearm, the bone projecting from the injury. With his wife's help, the Indian drew the injury back into as near normal shape as possible, and then cedar splints were laid on, plentifully smeared with spruce gum and then bound with watap or cord made from tamarack roots. Apart from the healing quality of the spruce essence, the gum hardened to make a perfect splint. When the Indian reached a doctor several weeks later, the arm had healed as well, the doctor admitted, as if it had been treated in a hospital.

The question as to whether the bears learned about spruce gum from the Indians or the Indians from the bears is easily answered. Bears have no means of communication with men. But men have a curious way of communicating with animals.

And they have found out, and still are finding out, all kinds of useful things.

The Hoist

For the benefit of newcomers to the hunting field, I have been asked to describe how a city slicker, weighing 150 pounds, some of it plain fat from living fifty weeks in town, hoists a buck weighing 200 pounds up into a tree.

For the buck must be hung up. To get the best venison, the buck must be gralloched, as they say in the Scottish deer forests, or gutted, as we say in plain Canadian, as soon after he falls as possible. The way you gut a 200-pound buck is the same way as you gut a one-pound fish: that is, you take its insides out. This can be done with the dagger that most new hunters carry on their belts, but it can be done a lot easier with a good jack knife which you find most guides carry in their pockets. No matter how well the beginner does this job, the guide will invariably do a little trimming after the buck is carried back to camp. And when you get the buck home, the family butcher will do still more, with many a grimace at the clumsiness of hunters and guides. But the main principle is, take out everything that will come. Same as a trout.

Now, with about forty pounds of the 200-pound buck removed from its weight, how do you get it up into the air and off the ground? And this you must do.

The first thing is to find the nearest limber sapling whose top, fifteen feet from the ground, will bend without breaking. You drag the buck to that sapling.

Next, with that dagger or that jack knife, you hack and whittle two good stout sapling poles which you judge will be strong enough, when joined to the sapling to make a tripod, to support the weight of the deer.

By this time, you will have blisters on your hands.

Now you take the piece of rope which you have unfailingly remembered to carry in your hunting coat pocket or tied around your waist. You tie it to the buck's horns, or, if you are extremely choosey, you tie it to the stick you have thrust through the gambrel joints of the buck's hind legs, and you shinny up the sapling, holding the other end of the rope.

The sapling will bend with your weight, providing you have not lost both wind and heart by this time. And hauling on the rope to get it as tight as possible between the buck's horns and the bending top of the sapling, you tie it fast.

Now, with the two poles (which must have forks or crotches on their ends, like old-fashioned clothes line poles) working them one at a time, you slowly and laboriously shove at the tripod of the sapling and the two poles, gaining a few inches at a time with the butts of the poles. And slowly the buck rises off the ground.

It is about the time you get the buck up into a sitting posture, with his hind end still on the ground, that you decide you are not going to give any of this venison away, not even to the boss back at the office. By the time you have the buck up so that only his feet are still touching, you wonder if duck shooting is not a better sport. Or even golf.

As a matter of fact, I weigh only 129½ pounds. And it has been my practice, over years of hunting, to use a far simpler method of hanging a buck. I just look around for the highest rock or hill nearby, go up there and holler and holler until somebody else comes, usually one of the guides.

That's the simplest trick of all, and I imagine a great many hunters employ it.

Power to the Unarmed

Game birds and game animals are supposed to be the concern of the sporting fraternity. Hunters take it for granted that deer, moose, bear and other wild creatures belong to them in some curious and special way. Year by year, however, the truth is slowly being revealed that non-hunting citizens, by far the majority, have an equal right in the game animals and game birds of our country with those who go hunting to kill them.

Provincial governments, watching with astonishment and even consternation the remarkable increase in the number of people taking up hunting as a recreation, are beginning to weigh the philosophy as well as the politics of the sport of hunting. Our pioneer generations across Canada come from countries in which game was jealously protected by the landowners and the common man had few if any rights in the matter of hunting for sport. When they came out here, game was one of the basic necessities of survival, and the tradition of freedom to hunt game was as jealously guarded as were the privileges of the landlords in the old countries. Several generations have since enjoyed this freedom.

But a new type of landlord is beginning to be heard from, that landlord being the non-hunting majority of those who have an equal right to the game and don't want to kill it.

Trigger-Happy

It is being predicted by sportsmen as well as by nature lovers that within ten or fifteen years, hunting licences will be issued only to experienced hunters with proof of that fact, and to members of recognized sportsmen's or naturalists' organizations that will certify the applicant has been a member in good standing long enough to have been indoctrinated with the principles of sportsmanship and a proper recognition of and respect for wild life.

John A. Livingston, executive director of the Audubon Society of Canada, reported a situation near a small thriving industrial Ontario city, which shall be nameless, in which there is a goodly population of vigorous men who like to take a gun out on weekends and hunt the neighbouring fields, woodlots and beaches. In this city also is a thriving naturalists' society which takes its pleasure in looking rather than killing.

On a survey of one half mile of beach near their community, they collected the following wild birds, all shot: 25 red-backed sandpipers, 10 pectoral sandpipers, 2 killdeer plover, 1 long-eared owl, 1 sanderling, 8 black-bellied plover, 25 horned grebes, 2 herring gulls, 3 ring-billed gulls, 1 Bonaparte gull, 1 great blue heron, 30 red-breasted mergansers, 1 hooded merganser.

"Can even the gunners," asked Mr. Livingston, "think of a reason why gun licences should be issued to people either so uninformed or so trigger-happy as that?"

The lid is slowly coming down.

Ducks' Peril — The Actuarial Facts

One of the officials of Ducks Unlimited in the Delta country of Manitoba points out one of the curious evasive attitudes of mind that are adopted even by people genuinely interested in the preservation of wildfowl. We flatter ourselves that the open season for ducks has been gradually reduced and narrowed down until the amount of shooting afforded sportsmen is surely at a safe minimum.

But due to the staggering of open seasons from far northern Canada down to the Gulf of Mexico, our wildfowl do not have a mere two weeks or a month of open season to face, but actually four months. The error in our thinking is that we put the emphasis on how much shooting there is for us. We should face the fact of the amount of shooting there is for the ducks. One-third of their lives they are being shot at.

How can it be otherwise? The sportsman and the needy Eskimo or settler-hunter in the far north must have his pot at the ducks as freely as must the sportsman from New Orleans and the puntman in the bayou of Louisiana, who likes a roast duck as much as any man. But the duck they all shoot at, born and raised for a few short weeks in northern Saskatchewan or the fringe of the barren lands, starts his journey in late August and, in easy stages, wends his way southward, to arrive in his winter home in December. And he is shot at all the way.

The fact, therefore, is that no matter how strictly an open season is narrowed in any one province or state, in accordance with federal control, and no matter how intelligently the season and the bag limit are accepted and enforced, the open season on ducks is four months whatever way we care to look at it. For one-third of their lives, year in and year out, the wildfowl of America are being shot at by ever increasing numbers of ever more mobile and ever more lethally equipped hunters.

Ducks Unlimited should, for psychological reasons, change its name to Ducks in Extremis.

An Irreducible Difference

A story that is enjoying a lot of popularity is the one about the lumber camp cook who, seeing a bear trying to break into the kitchen, heaved a stick of stove wood at the intruder. And the bear ran and picked up the stick and brought it back to the cook.

It was somebody's tame bear and all it wanted was a share of the amenities of a camp cook house.

When we befriend wild creatures, we usually seal their doom. In any summer, hundreds perhaps thousands of families have made pets of wild animals around their cottages. Raccoons, deer, foxes, wild ducks and many other pretty creatures have been fed and petted and invited to make free of the premises. They have thus been deprived of their greatest protection – their fear of the scent of man. The raccoon or the fox will walk cheerfully, even hopefully, into the first trap he meets this fall. The deer, scenting a man, will come expectantly into the open to greet a friend. The ducks that have been chummed with will be the first in the pot in October.

The hard thing to rationalize with regard to our wildlife is that those who would love to make a pet of a wild animal outnumber perhaps fifty or even a hundred to one the number who kill wild animals for profit or sport. In other words, we have a hundred men and women going about, with kindly hearts, making it easy for one man to kill.

Yet, you could prohibit man from hunting about as easily as you could prohibit men and women and children from loving wild creatures. In short, it can't be done. It stands as one of the great irreducible differences in human nature. And it is as old as that first cave man who, suddenly noticing a wild flower for the first time, sighed.

The Pelican

Bird watchers get a thrill, an aesthetic shock, out of encountering a bird that is far from its normal range and a record for their area.

But an Indian of the Calstock reservation in northern Ontario had a different reaction to an experience most bird watchers would have envied him.

He was out making an early season survey of his trapline when he saw, out on the water, the biggest wild goose he had ever imagined. The light was not good, the distance far, but the Indian was hungry, and a goose hangs no higher anywhere than in the opinion of an Indian. With his small rifle to took careful aim and hit the goose. Then he rigged a few logs into a raft and poled out to retrieve it.

It was a pelican.

"The Evil One," said the Indian, "has changed my goose into this ugly bird."

District Forester R. H. Hambly, reporting the case, describes the Indian's dismay as he hauled his prize aboard. To see a pelican at close range anywhere is slightly dismaying. The Indian looked down upon a huge bird with a heavy body, short stumpy legs and an enormous beak more than a foot long, very massive, with limp folds of a huge pouch on the underside of the beak.

After much deliberation, the Indian decided to bring the creature back to the reservation for his tribesmen and elders to examine, and perhaps to see in it some omen of things to come. What if all wild geese came to such a condition as this? Foresters heard of the bird and went to the reservation, where they identified it.

The white pelican is a well-known bird in the west, breeding from Canada all the way down to California. I have seen it in far northern lakes in Saskatchewan and throughout the prairies. Writing of it, P. A. Taverner, the famous Canadian ornithologist, said: "Pelicans are communists; individualism is unknown among them. The way one faces, they all face; as one poses, they all pose. Flying they assume their appointed positions, and, taking the beat from their leader, keep time with him, flapping and sailing all together. No more beautiful sight may be seen on the prairies than a long line of these great white birds, black pinioned, with golden pouches tucked under their chins. . . ."

But the one the Indian found in Northern Ontario was far off the beam, and gave the Indian a nasty turn, pinions, pouch and all.

Feeders

A good man among my acquaintances refuses to put out feeding trays and stations for the wild birds on the ground that it is cruel.

"The darn fools," he says, "should have gone south. If we put out food for them and detain them from their normal business of getting away south as fast as they know how, we are not being kind to them. We are seducing them."

The fact he overlooks is that most of the birds that come to winter feeding stations in cities and towns are already as far south as they are going. This is the banana belt to them. They will eke out a living somehow in the wintry gardens, parks and suburban fields. They are all around us anyway. And the purpose of putting out feeding trays is not so much to feed them as to attract them close, so that we may admire them or be amused by them.

The commonest feeding system is simply a low-sided wooden tray made out of any old box and nailed to a window sill. On it are spread such seeds as millet, sunflower, buckwheat, canary seed, and scratch feed. A dab of peanut butter in a small container such as the lid of a jar is a good attractor. The house sparrows will doubtless be the first to discover the treasure. They might even exhaust your patience and your expectations. But one fine day, a stranger will be among them.

You can saw a cocoanut in half and suspend the two halves, upside down, with a wire knotted through holes in them, from trees or porch eaves. No sparrow can feed from these upside-down feasts. But several species of winter birds can. And likely will.

If there is a tinker in the house, he can set up a six-foot pole in the yard, insert a length of quarter-inch iron rod in the top, and on that rod erect a small wooden house with a wind-vane like a ping-pong bat projecting from it, so that it swings in the wind, keeping the one open side of the house away from the gale. To such a feeding station come, in December, figments of last spring and of next.

Coals

After studying camp fires in peace and war for the past forty years, I offer impatience as the greatest bar to good camp cookery. In every fishing party, on every picnic, it should be somebody's pre-ordained duty to go ashore, well ahead, and build a fire and let it burn down to coals before even the tea pail is set to boil.

Flames won't cook anything. Red coals are man's oldest masterpiece.

Nine out of ten of the camp fires burning this minute, by the tens of thousands across the continent, are big leaping conflagrations. The tea pail is swinging in the weak yellow flames. On the lee side of the fire, the self-important cook is squatted, shoving the frying pan into the pallid excuse for a fire, dodging the smoke. After twenty minutes, the singed, smoked, sooty repast will be served. It serves them right.

An Indian and an Arab make a fire exactly alike. They make a very small fierce fire of the smallest, hardest wood they can find. They urge it on, until nothing is left but a bed of close, bright glaring coals. Then they start to cook. The Indian prefers a round bed of coals about the size of a frying pan. Over his, the tea pail rolls to a boil in four minutes. By that time, the cherry red is overcast with an ashy hue: and the fish in the frying pan sizzle oh, so gently. The meal is ready by the time you have washed your hands and taken one more look at the lovely land ahead.

The Arab likes a long narrow fire, but not much larger, in area, than the frying pan fire of the Indian. Over this, he extends a steel spike or skewer, three feet long, on which are impaled in turn: a bite of lamb, about the size of a fifty cent piece, a bite of onion, a bite of fruit – whatever's handy, a bite of lamb, a bite of onion . . . He turns his hand slowly, as the impaled bites hiss.

But the coals must be red, tender, bloomed with ashy grey. And the skewer held close.

Restored Silence

As the freeze-up approaches all across Canada, peace is descending on the wilds; the hunters have come to or near the end of their open seasons; only the hardiest gunners in a few favoured regions are still facing the weather in pursuit of their game. What an eerie business it must be for a wild creature to comprehend life. For nearly ten months of the year, they dwell in security. Not merely security – for those that come in contact with the curious two-legged monster with all his unearthly powers on land and water find him aggressively friendly, offering gifts, begging the wild things to come near, pointing cameras, leading little children close to behold the beauty of the children of the wild.

Then, without warning, suddenly all is changed. In an instant, every marsh blazes with furious guns. Through autumn woods, the two-legged monster, so lately kind and softly enticing, stalks warily, lightning in his hands.

Gone are the roadside visitors, with their popcorn, apples, peanuts: and when a deer sees a man, it is death he beholds.

All over the wilderness, from the Arctic goose marshes down through the moose's muskegs, the deer's hardwood hills, the partridge's gullies, the ducks' ponds and lee shores, every wild creature runs, hides, flies high, flares wide, in a universal fright and unbelief.

Then, in an instant as inexplicable to the creature mind as the instant on which it all began, in September, it ends. Not a shot is fired. Out of the wilderness troop the hundreds of thousands of hunters, giving it back its silence and its proper mystery.

I do not suppose the wild creatures have any understanding of mysteries. To them, life itself is without explanation.

But it must be strange to be a wild creature, now that peace has come again in the wilderness, and to listen to the emptiness.

Nature's Rights

In Britain they have the National Trust which takes over and administers historic sites, buildings and a great many other items of value to the future which might otherwise be swallowed up in the often cheap flash floods of progress. They also have the Nature Conservancy which has taken over a number of areas characteristic of the different biological regions of Britain. And these areas are not, as one might expect in this country, for the benefit of the tourist trade. In fact nobody, not even scientists, can get into a Nature Conservancy enclosure without the greatest red tape and rigamarole. I asked a noted British biologist what was the purpose of such rigid enclosures. "Simply the belated recognition of the fact that natural creation of plant and animal has a right to co-exist with us for no sake but its own."

Rather quietly, the Nature Conservancy of Canada has been founded. It will be interesting to watch what fortune it has against the massive new-world forces, tourism, mineral, forest and water exploitation and the ever divine right to make a buck. So far, on this continent, nature has no rights in the granite face of man.

Yet who but nature assigns human rights?

The question of human rights requires some thoughtful consideration. Nature assures rights to no creature, not even man. And with the world shaping up the way it is, with its enormous stockpiles of armament beyond the wildest dreams of Genghis Khan or Napoleon on one hand, and the steadily seething populations of the Orient and Africa threatening to spill out in all directions on the other, the thought can hardly be suppressed in the minds of those who study nature without prejudice that nature may be up to one of her tricks. In the aeons of time that nature has been governing the earth, directing the rise and fall of animate empires, rights have been accorded to no species. Great giants, vaster than man, for a million years have trod the earth, mastering it. And just about the time they assumed their rights, they were quietly interred in the badlands of Alberta.

Rights, though it is a six-letter word, is a naughty word to use in the hearing of nature. Maybe nature is not so compassionate and humorous an old lady as we have been led to believe.

A famous French biologist, an expert on famine in the Far East, told me at the time when there were early rumours of an impending devastating population explosion, "You take consolation in the fact that in the long past, many forms of life have tried, or been moved to try, to take over the earth completely, in the fashion that our human species is now trying to do. But nature arranges it for such species as get out of hand to attend to their own destruction. Maybe they consume themselves out of existence. Maybe they are so efficient in one environment that when the environment changes, they are helpless in the softness of their complete efficiency. On the strength of past performance, I am perfectly confident that nature can handle our species quite as effectively as she did our predecessors. Gentlemen, we have nothing to fear. Nature will beat man in the long run."

There are both philosophers and scientists who are perfectly convinced that nature one of these days will take us in hand.

In what way will nature put us in balance with the rest of creation?

WINTER

The Naked Forest

If, on a bitter December day, we pity ourselves in our discomfort having to walk a couple of blocks with the sleet in our faces, or having to stand huddled for six minutes awaiting a delayed bus, it is helpful to think for a minute about the wilds at this same hour. All over the continent, in areas vastly greater than those square miles we humans have occupied, there are creatures of every size and shape to the number of so many more millions than us that nobody would dare to estimate their total number.

And at this unholy moment of cold and gale, they too are engaged in the business of keeping alive. In the tangled swamps, the moose; on the mountains, the goats and sheep. Across the marshes, through the forests, the wolves, deer, foxes. In the thickets, the hares and rabbits; in the fields and barrens, the mice, voles, shrews, lemmings and a hundred creatures you have never heard of and are never likely to see. Millions upon millions of them, sharing with us the mystery of a thing called life, or, shall we say, living.

The sheltering foliage of the forests and thickets is gone. The escape routes of stream and pond are locked. Before the heavy snows, everything is open and bare to the elements and the enemies. Yet by a curious dispensation of creation, these millions of our fellow-travellers are spared the gift of self-pity.

Snow

If it were not for the shelter of the fallen leaves, treacherous and tale-telling as it is, the foxes and owls, being very human, would soon destroy themselves by wiping out their means of subsistence.

But, very hushed and still, amid the rustling and quivering dead leaves, the mice, shrews and rabbits lie low for a few brief weeks in order to guarantee the survival of the foxes and owls. Then comes the snow: and all is well. The hare, turned white, is invisible. The cottontail's bomb shelter under a brush pile is insulated and made almost impenetrable by a batting of snow. And the mice go subterranean. Along the fences, under the tiny arches of the buried leaves and shrubbery, they have their underground system as good as any city's.

The ruffed grouse, instead of spending the night in some bare tree or in the doubtful shelter of a spruce or cedar, dives into a snowbank, probably scaring the daylights out of many a mouse by such dive-bombing tactics.

By hard work and vigilance, the owl and the fox and the rest of them are able to find enough of these snow-hidden creatures that eat vegetables: and so life is continued. But always on a business-like basis of scarcity. Nature never seems to let anybody have an easy time, except the porcupine.

The greatest of all snow sights to see is an otter hunting for mice or other provender under the snow. He dives into the snow as he dives into water; swims along under the snow for long distances, exploring the tunnels and dug-outs; then emerges bursting, somehow hilariously, out of the snow for a look about him.

In nature, there is no real security.

The Constant Mate

Foresters in the north of Ontario reported that they had found two Canada geese in an open stretch of a river during the first week of December. This was several weeks after the wild geese had left for the south.

"We were able to get within a few feet of them in the outboard skiff," said the foresters, "before the geese took wing, and then they only flew a hundred yards or so before alighting again."

What the foresters were witnessing is one of the great romances of wild life. The Canada goose mates for life, and rarely if ever does one goose select a new mate if the first is lost. What was probable in the case the foresters encountered was that one of the pair had either been injured or had grown too old to undertake the journey with the flock. So there the two of them were, sequestered in a lonely northern river soon to be bound in ice, and awaiting in unquestioning companionship the fate nature affords her own.

Chipmunk

Human beings manage to think well of their own species despite their continuous failure to make themselves comfortable. For several thousand years now, they have been writing books in their own praise, inventing songs for their own glorification and making endless speeches of congratulation to their own kind.

Actually, they have no sooner got one little scheme worked out for their own security and comfort than they think up something new to destroy it. Their latest masterpiece of folly is the building of magnificent cities with every conceivable convenience for comfort, and then rendering the cities uninhabitable with traffic.

Chipmunks, on the other hand, worked out a system of civilization of their own countless ages ago, and are sticking to it famously, without having uttered a single sound of self-praise.

By the end of November all over the country, except on the Pacific coast, these model Canadians have retired for the winter in far greater style and comfort than those who can afford to go to Florida until next April. While we are busy all summer racing around in traffic, each chipmunk, entirely on his own, was burrowing a hole about one and a half inches in diameter, straight down into the good warm earth for about three to five feet. From there, he levelled out and drove his tunnel, zig-zagging this way and that, for as much as ten to twenty feet in length. Off the tunnel he constructed, at pleasant distances, rooms from the size of a mug to the size of a six-quart pail. These he packed with seeds, acorns, dried berries, the odd June bug and other comestibles, insulating them all tidily with cut leaves. He even built a small room for a privy. He might have eked out his energy by creating a couple of spare exits. All exits he carefully closed up before the first hard frosts. And now he is rolled up in a warm ball in the best chamber, having wished us all a soldier's farewell until spring. Mostly, he will sleep. But now and then, he will wake drowsily for a nip of June bug or a snack of wintergreen berry. Yet without a trace of vainglory!

Muskrat

The coming of ice to pond and stream is a godsend to many creatures, the muskrat most of all. It brings him peace. A muskrat has a host of enemies, not only on the ground and in the air but under the water. Rabbits are supposed to be the bread of life to the wilds. All the animals that hunt rabbits also hunt muskrats with zeal. Wolf, coyote, fox, hawk, owl count on so many muskrat dinners per season. But in the water, the muskrat's deadliest enemy is the mink. And big snapping turtles, pike, muskies, and even big-mouth bass, when the rats are young, are ever-watching from below. A muskrat doubtless envies a rabbit its security.

But when the ice comes, the muskrat dens in the mud banks and the muskrat houses out in the swamps are fortified as with concrete. The air-borne enemies are fewer, the land enemies are more easily detected, and the underwater enemies, all save the mink, go away somewhere to doze for the winter.

The muskrat does not store up winter food the way the beaver does. He still has to go out each day for his fill of whatever he can dig underwater in the way of roots, clams or crawfish and whatever he can find along shore near the open water he must always have near his home. Rations may be lean in winter. But until another enemy, man, comes rubber-booting through the slush on the verge of spring, the muskrat has a comparative season of freedom from fear.

The least of his enemies, man, takes about three and a half million muskrats a year in Canada alone. But the muskrat is a rat, in full fact. He fights like a rat, travels and explores like a rat and therefore survives like a rat. If it were not for his enemies, there might be too many of him.

Blue Jay

A considerable number of us are like blue jays. We resent being shown consideration.

When you put out food for the birds in winter, all species save one appear to be grateful. If there are any loose pigeons around the neighbourhood, they take it for granted you are thinking of them first and foremost, for they come enthusiastically to the feast and try to get all before even the starlings wake up.

The starlings and sparrows are about evenly matched. There is a humble, slightly incredulous air about them as they arrive, hasty but circumspect, at the outskirts of the feast, and they hesitate a little before intruding on the pigeons. If you try to shoo off the pigeons, it is the starlings and sparrows who assume you are referring to them.

The wild birds, the chickadees, nuthatches and other precious wanderers of winter, come to the feeding station, even for the hundredth time, as if it were something beyond comprehension, another of the acts of God.

But the blue jay is a character. He doesn't want to accept a gift. It goes against the grain. If he can't steal it, he doesn't want it. Thus he flits furtively around the neighbourhood, uttering clandestine calls, and darts in to grab a peanut in definance of you and all generosity. And when he reaches a safe distance, he gives a strident cry of derision. He has outwitted you.

Pauses

In talking to an Indian back in the bush, an Indian not accustomed to our white way of conversation, you have to allow plenty of long, empty pauses.

We white folk hate empty pauses. It may be the radio has trained us to attempt to fill every second with chatter, otherwise we feel the time is lost. Thus, if the party we are conversing with does not instantly jump into each pause we allow, why, we jump in ourselves.

Indians are not like that. The pauses in their conversation are just as important, just as precious with meaning, as the spoken passages. Up in James Bay I was asking a Cree what the curious tripods were that we came upon along the lonely shores of ice, tripods four or five feet high, and set a hundred yards or so from the willow-clad shore line. Owl traps, said the Cree. There is a small muskrat trap set on each tripod, and the big snowy, or sometimes the great grey owl, seeing a handy perch, drop down and are caught.

"What for?" I checked.

"To eat," said the Cree.

"Are they good?" I asked hoarsely.

"Very good," said the Cree.

"A white man once told me," I submitted, "that they taste of mice."

"Yes, they do," agreed the Cree.

And here is one of the places you leave a pause in conversing with an Indian. I left the pause. The Cree stood reflecting. He was searching his mind for some way of explaining to a white man about eating owls.

After a long pause, in which I had time to think of a great many things, my mind darting this way and that erratically, in the white fashion, the Cree said:

"Mice are too small to catch. They are too little. You would need a lot of them. So the owl catches the mice and eats them. Then we catch the owls and eat them. That is how it works."

Perfect economics. Of course, it glosses over the little matter of flavour, but in economics, flavour does not count. In the study of economics, Indian or white, it pays to leave some pauses for an answer.

Thought for January

However bad our luck may be in the onrushing season, a certain number of fish are bound to land in our creel. They are fated. They are our fish already, although they do not know it.

There, in the dark, icy pools they lurk, this minute, awaiting their doom, so curiously knit with our luck.

If they are trout, they have dropped back from the swift riffles and fast waters where, soon, we are destined to meet them, and are lying in the deepest pools, under the perilous ice. Many of them have retreated a mile, or five miles, to some lake, where, under the massive lid of ice, they explore the shadows, picking up what food offers in that half-life or semi-consciousness which is the lot of many fresh-water fishes in winter. But they are our fish, in whatever gloom they lurk. Kismet. It is fated.

If they are bass or pike, they linger, semi-torpid, amid the shadows, clefts and crevices, their appetites so suspended by winter that only rarely do they feel inclined to move at all in quest of a small hors d'oeuvre of a semi-somnolent minnow.

Hundreds of miles far to sea, sweeping the vast spaces of the ocean the way symphonic chords sweep the air, the salmon, all unaware, are moving our way – to us, individually, you, me and whatever other Joe is fated with them.

Pecking Order

Crows have not many friends or acquaintances in the bird world. As a rule, when a crow appears, smaller birds either take cover, raise a fuss or, like the kingbird, launch an immediate aerial attack on the black intruder. It was with considerable surprise that I witnessed, just before the ice went out of the northern lakes, a couple of bald eagles behaving very chummy with several crows.

On the ice lay a carcass of a skinned wolf that my trapper host had set out as a feast for the eagles. These birds spend the winter on the Great Lakes, about twenty miles from the neighbourhood. They subsist on fish which are thrown out of the nets of the commercial fishermen along the shore, but come inland from time to time looking for other carrion. The skinned wolf became an all-day lunch counter for about thirty herring gulls, half a dozen ravens and four crows. And whenever the eagles appeared, the lesser birds scattered.

The eagles feasted one at a time. Both were in the adult plumage of white head and tail. One invariably landed on the ice immediately beside the carcass. The gulls, ravens and crows flew about widely. The second eagle settled about twenty feet away. Then the lesser birds landed in a polite circle about forty feet from the carcass. At the slightest attempts of the second eagle to waddle in a little closer to where its mate was eating, the bird at the carcass half raised his wings, bared his massive beak and threatened his mate. When he had finished, he would fly off to a dead tree and perch. Then the second eagle hastened to the feast.

But with both eagles, the ravens and crows were equally welcome to fly in, after the preliminary scatter action, and, working at the far end of the carcass from the eagle, to feed in perfect safety and security. They were not five feet from that massive beak and those powerful talons.

But never a gull was permitted within less than forty

feet. That was the radius of the circle the gulls kept, wailing and hollering their grief without ceasing.

Endurance

What goes on in the wilds, during the months of winter isolation for vast areas of our northern expanse, is becoming less of a mystery, now that government biologists are being employed to find out, and have the aid of air patrols to study animal movements.

Out from Gogama, in the northern part of Ontario, the rangers spotted the heavy ploughed track of a moose doing some unseasonable travelling. Moose usually hole up in deep winter. On coming down for a clue, the rangers saw wolf tracks along with the characteristic moose furrow. They followed by air to where the chase entered a swamp.

Landing the plane, the rangers followed on snowshoes. A little way in the swamp they found where the pack of eight wolves had given up the chase and had lain down, obviously for some time, and rested.

Then the wolf tracks quit the moose trail, and headed out of the swamp. The rangers followed up the moose track; and within 150 yards of where the wolves had given up, they jumped the moose from its bed, where it has been resting.

Eight wolves against one moose. On its long legs in the soft snow, the moose had dead-beaten the wolves that hadn't enough heart left to go 150 yards farther.

Hibernation—Real and False

Despite the popular belief, bears do not hibernate in the true sense of the word. They merely go into a darn good sleep like a fat man. They don't even find caves or dens, as folklore has it. A bear will snuggle down under a cedar tree or behind some old roots whenever he feels that good old fat man's drowsiness stealing over him. And then let'er blow and cover him up. There he will slumber all winter through, if undisturbed.

But if you poke him, he will come out fighting and wide awake.

True hibernation is experienced by the great family of ground squirrels such as the gopher. Dr. Otis Wade, outstanding American biologist, found that the striped ground squirrel, which has a normal temperature of around 105 degrees, drops, in the depth of winter in his burrow, to a body temperature of as low as thirty-six degrees; and his active heart-beat of 200 to 350 a minute falls to five a minute. It loses as much as forty per cent of its weight by April. Dug up and held in the hand, it seems dead. A bat, which is supposed to hibernate, if held in the warm hand for a few minutes, will come to life and fly around.

Grizzlies and the great Kodiak and other northern bears den up a little deeper than do the black bears, in rock slides and caves if possible. But the polar bears have their own system. The females assemble in certain gathering places, certain known islands and peninsulas of the Arctic to den up and have their young. The males are believed to spend the whole winter on the prowl. Bears love sleep; but they are nonetheless as wide awake as the most high-strung of animals.

The Hungry Fox

If in February we are yearning for the first signs and portents of spring, imagine the yearning of a fox or a coyote or any of the other animals condemned to staying awake all winter.

February is the leanest season of the year. All the easy-to-catch mice, voles, lemming mice and other small bread of the field and forest have long ago been caught. What remains of small eatables is very wise, experienced and alert. All the foolish grouse have been consumed. Only the wise grouse remain. Every nook and cranny where something to fill a lean belly might be hidden has been scoured.

To be a meat-eater in mid-February is a tough assignment. But to be a fox is tougher still. For this is the mating season for the foxes. You can hear their weird little barking in the night. It is sad to be both hungry and in love.

On those quadrennial years when the great snowy owl comes down in large numbers from the Arctic the problems of the fox are further complicated. Every fourth year the lemming supply in the Arctic runs out; so down float the lovely white owls to feed in these more plentiful climes for two or three months. Doubtless other owls move southward, too. The mice, the cottontails and swamp hare that, for three years, have devoted their Februarys to watching and listening for foxes this fourth year must wear eyes higher on their heads than ever; the air is full of death for them. The snowy owl and the short-eared both hunt by day as well as by night. The great horned owl and the barred owl and all the small owls prowl in darkness.

So does the fox, poor chap, the hungry fox hunting for a hungry mate, and any little thing to eat that might offer, en route. If his little bark in the night sounds querulous, you can understand why.

Faith

The day before a team of husky dogs in the north has to make a journey, the dogs are never fed. They are fed the next day, at the end of the journey.

This policy has two angles. The unfed dogs are more spry and agile. But mainly they are inspired by the knowledge that they will be fed when they get to their destination. And such is dog psychology, they pull like devils in the knowledge that they will eat at the end of their run.

Food is a dog's religion. When you see a dog sitting rapt and spellbound watching you eat at table, you are beholding a dog in the act of worship. By the intelligent application of this fact, you can train a dog better than with a whip. Most dogs are fed once a day only, usually at supper time. The hour before they are fed is the time to undertake to teach them anything from manners to tricks. At that hour, they are expectant and alert. After three or four such lesson periods, it begins to dawn on them that they get fed as soon as they comply with some dopey notion the master has in mind. You cannot teach a fat dog is an old saying. It is hard to teach a fed dog. Hungry dogs are not eager to learn. They are merely hungry to eat.

The Indian who last had me for a passenger told me he had some small harnesses which he rigged on the pups some weeks or months before they were ever entrusted with a place on the team. When they were old enough to travel, they came along in their dummy harness, though not pulling. They too were starved the day before, and got fed at the journey's end. When it became their time to become sleigh dogs, they were already indoctrinated. Like their elders, the hunrgy pups were almost eager for the harness, knowing that harness meant food. It is not such a bad religion. Countless human beings are activated by the same faith.

Beaver Skin

In the James Bay country there is a tribe of Indians, probably Crees, who turn out a beaver pelt that commands from twenty to thirty per cent more from the fur buyers than the standard hides in the fur trade.

The ordinary way of preparing a beaver pelt is to scrape all flesh and fat off the hide, being careful not to injure or cut the fibre of the skin, and then stretch it as tight as a drum head on a hoop. It then dries, and is said to be cured.

These Crees stretch the hides, but leave them out to freeze, bringing them in every few days to be washed with some secret and traditional soap mixture and further scraped and kneaded while banjo-tight. When the fur dealer gets the pelts in spring, the skin is as soft as chamois.

The Crees assert that that is the traditional method of curing beaver furs from time immemorial. We often wonder how the Indians kept warm in winter in pre-invasion times. Imagining them dressed in buckskin without underwear on a frigid day is enough to bring us out in goose flesh. Jacques Cartier found them clothed in undergarments of smoke-tanned deerskin, but with magnificent robes or overcoats of beaver furs. Every Indian, man, woman and child, had a beaver robe, its hide as soft and clinging as chamois, to wear wrapped around from head to foot; and in their homes, they had piles of beaver robes to sleep in.

The beaver, of course, was everywhere in those days, and one of the easiest furs for the Indian to obtain. It was the sight of these glorious robes that inspired the idea of the fur trade in the early invaders who were trying to get round to India to locate spices, gold and other trinkets.

If anything goes wrong with civilization, these Crees up around James Bay are going to be all right.

Deep Freeze

About mid-February, the fatal thaws occur which are the harvest home for the wolves. While the snow is deep and soft, a wolf is on short rations, snuffling out small game like mice from their tiny dens. And poor provender that is for a big, winter-lean wolf.

But after a thaw, on his broad paws, a wolf can make speed and cover large areas of country over the crusted snow. It is the deer he seeks: the sharp-hooved, spindle-legged deer, gathered in their winter yards. And the bucks have lost their antlers.

Anybody who has ever seen a deer yard after a family of six or seven wolves has raided it can hardly help but be seized by a permanent hatred of wolves. On the east shore of Lake Superior, an Indian took me into such a yard. In the frozen cedar swamp, seven deer lay dead among the paths they had kept open all winter amidst their hide-away. Only one had been really eaten. The other six lay torn only enough to be killed, with a small portion chewed in their flanks, as though the wolves had wanted no more than a tid-bit of the kidneys. We trailed them on to a distant swamp where they had indulged in another slaughter.

The Indian did not share my fury.

"Plenty of deer got away," he said calmly. "In maybe six weeks, female wolf dens up in hills near these swamps to have pups. The female come down and feed on these deer. Pups feed on them too. . . ."

"You suggest," I protested, "that the wolves deliberately slaughter these deer and put them in storage for the whelping. . . ?"

"Yes," said the Indian. "They bite that hole in the flank so deer won't bloat and spoil. Freeze better. Keep longer. But of course wolf likes bad meat. . . ."

At all events, Indians, wolves and deer got along all right in the ages before the white man came to moralize upon them.

Night Raiders

The bush wolf, as the coyote or prairie wolf is called in the east, has penetrated far in off the prairies through Ontario and Quebec, thanks to the clearing of the land and the construction of roads and railways. A doctor who owns a sort of recreational farm twenty miles outside the city limits of Toronto had his pheasant coop raided by some animal that dug its way expertly under the buried edge of the chicken wire. Eleven pheasants were killed. The tracks appeared to be a dog's. The doctor got his saddle horse and trailed the animal in the snow back through a woodlot and was astonished to jump a fat, well-fed coyote. He had no gun.

So he set several jump traps at the excavation under his pheasant coop. He did not catch the coyote, which was too smart to return, like a murderer, to the scene of his crime. But what the doctor did catch were: one fox, one red-tailed hawk, one goshawk, and five great horned owls. All on the one setting. And all the creatures, even the big owls and noble goshawk, had tried to foot their way down through the tunnel dug by the coyote.

Little do we realize, says the doctor, what goes on in the night, even on the fringes of great crowded cities.

Of course, a pen of grain-fed pheasants would attract quite a company of speculators from amongst winter-hungry wildlife. Perhaps for miles around. Pheasants may smell as good raw as they do cooked.

Grace

In the army, the prescribed grace before meals is a perfunctory, "For what we are about to receive, thank God!" Even the chaplain, if he is present, generally confines himself to this abrupt address to the Deity.

In the first war, I had an invitation to the mess of a general who was known to be a deeply religious man. He did not leave the matter to the chaplain, the president or the vice-president of the mess. As we stood to our chairs, the general said: "Gentlemen, we will each ask a blessing." And the whole tableful of us stood with heads bowed, doing a private job, each in his own fashion. In some sects, in many homes, this same principle is observed. Nobody speaks. Each member of the family, at a signal from Dad, bows the head and asks grace in a personal, not a perfunctory fashion.

On the street I met an elderly and prosperous lawyer with a large and heavy paper bag.

"My grace," he explained, indicating the bag.

"Ten pounds of seed," he elaborated. "I've just come from a seed store. Three pounds of millet, three pounds of scratch feed, a pound of sunflower, a pound of buckwheat, a pound of straight canary and a pound of rape. That's my grace before meat."

Wild bird food.

"I have a feeding tray on my dining room window sill," he said. "And another under the shelter of a rose arbor in the garden. Every day I say grace by going out and sprinkling a tumbler full of these seeds for the birds. Every Sunday, I add a special grace by putting a hunk of suet about the size of my fist in one of those open-work wire soap dishes nailed flat against the tree."

"And," I enquired, "do you say grace while going through this ritual?"

"What's more," replied the lawyer, "I feel it."

Crossbills

I was sitting on a snowy expanse of rock, with spruce around, watching for a family of otters that were expected to come asliding to a slope of rock I had my glasses on. It was one of those keen, still afternoons, following snow.

Suddenly I was conscious of the air right around me being filled with a tiny tempest of bird sounds, chirps, hard little chips and jeets, it seemed like hundreds of them all at once.

On turning, I got the shock of my life. Somebody there in that wilderness place had set up a Christmas tree right behind me! You have doubtless experienced that reeling sense of unreality in the woods when something magical appears. This Christmas tree was a spruce about twenty feet tall, and it was spangled, festooned and glittering with a hundred or more bobbing lights, exactly like those Christmas baubles, red, dark red, green, orange and red, red and green. . . . It was, of course, a flock of red crossbills about a hundred and fifty strong that had swept over the rocky plain and landed, all of a cluster, on the one tree.

Rhubarb, that curious conflict of red and green you see in a rhubarb stalk, is the colour the red crossbill suggests, the males being an eye-aching brick red, the immature males various shades approaching orange, with dark green suggested in the wings, and the females grey with a yellow almost as rich, on a snowy tree, as the red of the males.

Upside-down, any old way, they bob and swing on the spruce and hemlock cones, the size of sparrows, but the colour of Christmas, or of an antique king's raiment.

The Bath

We know that there are millions of birds wintering even in the northern half of America, despite our delusion that all the birds are away in the tropical south. Around our major cities, where bands of expert bird census-takers could make a reliable count, as many as fifty to seventy species of wild birds are found cheerfully residing with us. How do these birds fare on a night of savage sleet and frigid rain? Well, the experts say that many do perish. But nature is not so savage as she might appear. A bird's temperature goes as high as 112 degrees. It has the most fantastic insulating system in its feathers, by which that high temperature is conserved, even in the most intense cold.

One frigid day in January children in the adjoining yard to mine were flooding a backyard rink. Ice was forming almost as fast as the spray fell from the hoses.

And to the astonishment and mild horror of us all, a company of house sparrows, starlings and pigeons gathered to dive down into the spray, shake their feathers enthusiastically to get the water into them for a good bath. And, well wetted, they flew to the fences to flutter themselves dry. And warm. And in no time were chirping and cooing and chattering as comfortable as could be.

In nature, we're the big softies.

Bird Tourists

With binoculars around my neck I completed a four-thousand-mile winter motor trip down through Kentucky, Mississippi, Alabama to the Gulf, around the Gulf and up through Texas to New Mexico and Arizona. I saw thousands of our birds on their winter vacation – robins, bluebirds, redwings, grackles, hawks of every description, herons. And the important thing I have to report is that they do not feel at home down there. They are distinctly tourists. They might just as well have had binoculars around their necks for looking at me.

When the birds are on their home ground in Canada, they have an air of attachment, of belonging. They are busy. They have nest sites to select, and nests to build. They have territories to declare and to defend against others of their own kind. Each bird, when he arrives back home here, adopts a proprietorial and homelike air.

This is entirely missing in them down in the south. It is the first thing to strike you about their behaviour, even though you see them but in passing. They are aimless, wandering. It is a country simply teeming with food for them. The climate is mild; more than half the landscape is green from the evergreen live oaks and pines; the streams run limpidly, insects hum in the February air, and flowers of some kind are to be seen at every glance. In the Gulf states is enough space and food for all the birds in America. But of course millions of them went on down over the equator to summer.

But it is not home to them, in either place. It is merely a winter resort, balmy, lazy, abundant in all things. And the birds act precisely as do those of us who go down for an escape from winter. They sit in the sun, with an indolent and slightly dissatisfied air. They move about from one area to another as we do, for no reason but for a change. At home here, a robin wears an air and attitude of purpose that few other birds can equal. He is a person of affairs.

Down in Louisiana, a robin cuts the most idle, dejected figure you can imagine. At home he is forever alarmed. Down south, nothing alarms him. If you shy a divot at him to stir him up, he flies off without direction or purpose. He never utters a sound. He is just an idler, a bum, killing time.

Winter Bird Spotting

The reason so few of the public know about winter bird watching is that so few people get up in time Saturday and Sunday morning. Hardly anybody sees us, dressed in our parkas and hunting boots, leaving the house at 7 a.m. or earlier, to drive out to the country to check birds.

Coasting along back winter roads, tramping through woods and woodlots and frozen swamps, we search for birds, our binoculars on our chests. You may think this is a pretty small game: a bird identified. But we may, in a lovely winter morning's tramp, locate and identify thirty-one species of wild bird, from a ruffed grouse to a chickadee. One by one, we score them for our day's tally. And then, out of the blue, we flush a bird we have never seen before, summer or winter, anywhere. It may be a rare visitor. It may be a wanderer far off its normal sphere. But for us, it is a record. We add it to our winter life list of unseen things seen.

Dry Boots

A fur trapper friend of mine was walking ahead of me on the ice, visiting beaver traps. The beaver house we were approaching was in a marsh through which a creek wended; and the ice was tricky.

The trapper, with a wooden-handled ice chisel five feet long, was tapping the ice ahead of him, to test it. Suddenly, without warning, he went through the ice up to his middle.

With a scramble, he slid himself out on solid ice and began rolling frantically over and over in the snow, sliding his legs into the snow, beating them with his hands, heaping snow on them, as he twisted and turned.

Then, calmly, he stood up, and continued to pack snow to his knees and thighs, beating it off, packing it on, beating it off. . . .

"Not a drop in my boots," he announced cheerily.

The dry, crisp snow was a regular blotting paper. It instantly absorbed the water. By fast action, the trapper had drawn the icy water off and out of his trousers and sock tops before much of it had penetrated to his underwear and before any of it had got down into his snug-laced, leather-topped hunting rubbers. It was a wonderful demonstration of the sort of speedy reaction to violence which is part of the daily lives of a million Canadians who dwell outside the fringe of our familiar security. In the wilderness, men simply perish if, in the first place, they do not know the lore of the wilds; and, in the second place, if they are not almost automatically adjusted to it. I do not suppose three seconds elapsed from the instant the ice broke until the trapper was rolling in the snow like a maniac.

Poison

When a leopard escaped from a zoo, the trick that caught him was drugged meat. A semi-conscious leopard was easily restored to his cage. A few years ago, so the legend goes, big game hunters in Africa used to place drugged meat near the waterholes for the lions to find. And guides, watching from a safe distance, signalled the sportsmen to come on in and shoot, after the lions were drowsy or asleep. There are various ways of being a man's man.

The use of poison against animals is now outlawed nearly everywhere on this continent. Until a few years ago, the use of strychnine against wolves was permitted. Fur trappers like to rid their territory of wolves not only for the bounty, but because the wolves destroy a lot of fur in the traps.

On a flight into a strange section of northern Canada, one of our party was a very skilled trapper and woodsman. The first day of camp, this man went for a tramp on snowshoes and came back to report:

"This territory has been poisoned over. There isn't a track of any kind to be found."

Subsequent enquiry proved that three years earlier, a lawless trapper had used poison thereabouts.

The trapper poisons a deer carcass left by the wolves, or else deliberately shot by him for bait. Whether wolves get it or not, other creatures do, large and small. The fox, raven, Canada jay, eagle, mink, lynx, martin, fisher, if they do not eat of the original bait, find and eat those animals who have. Death and destruction, travelling on the wings of birds, in the bodies of mice, fox or mink, spread far and wide, over miles of territory. It takes nature a long time to repopulate a poisoned region. A good woodsman can detect a poison area in the trackless snow.

Snowshoe Hare—Bread of the Woods

We do not usually think of bunny rabbits as meat-eaters. But the snowshoe hare or varying hare, which, in a large part of Canada is that attractive little denizen of cedar and spruce swamps that changes its coat each year from bush brown in summer to snow white in winter, is declared by trappers, and scientists who now confirm it, to be fond of frozen meat.

Trappers who bait their traplines with meat, often with snowshoe hare meat, have actually caught the gentle little storybook animal red-handed, or red-toothed, in the act of feeding on the frozen meat. W. H. Burt, in his book on the mammals of the Great Lakes region, states that trappers claim the bunnies will not touch carcasses that have not been torn or bloodied. In other words, they will not damage trapped dead animals, so destroying pelts, as do several other animals that disturb the traplines. But there is no doubt they will eat frozen flesh if it is exposed in the bait.

There must, of course, be times when the varying hare will eat anything. For with the exception of the lemming this animal is the most sensational exemplar of the cycle principle in wild life. In a cycle of nine to eleven years, their numbers will increase from one per square mile to several hundred in the same area. During these peaks of population, when their cedar swamps become so crowded that the hares range out into the woodlands, the competition for food during the height of winter must be very severe.

The "crash," as the biologists call the sudden fall-off of these excessive populations, is one of the most spectacular events in natural history. Predators move in to the crowded area. Disease seems to spread like fire. Other causes not yet determined by science seem to be at work. And in a single season the hundreds are reduced to one per square mile again. And the slow cycle of rebuilding starts all over.

In the night, I heard unearthly screeches coming from some trees, and when I shone the flashlight beam, I spotlighted a great horned owl with a showshoe hare in its clutches. The rabbit was making real meat-eater sounds, and I was a little sorry it did not take a bite out of the owl, to see what owl meat tastes like.

But it seems the hares have a due sense of fate, and are humble in its presence, except for a few screams. They appear to know they are what naturalists have called the "bread of the woods."

Rough Leg

Where the eight-lane super highway approaches the big city through the stark factories and plants far-flung and crouched across the wintry landscape, a rough-legged hawk hovers. A hundred yards away, the grim towers of electric transmission lines stand stiff. Less than a mile distant, an international airport spouts up its jets or blinks them howling down. On the eight-lane highway the late afternoon traffic roars darkly, endlessly. And there, fifty feet up, its eye on a patch of wintry weed stubble by a factory, the rough leg beats its wings skilfully, holding itself poised for the stoop.

The rough leg – not the ferruginous relative that lives on the prairies but the common rough leg that breeds in the barren lands across the top of the continent – never sees men up there in his native land. He feeds on the lemmings and other small rodents that abound on the tundra. And when he drifts south in winter, he finds good mouse and rat hunting in the immediate vicinity of a very curious creature that lives in a state of stupendous frenzy, noise, fume, and senseless activity. With beautiful fierce eyes he concentrates on mice, and is totally unaware of men.

Silent Watcher

Nearly seventy-five years ago, Ernest Thompson Seton wrote: "The silent watcher sees most." He was referring to those who take pleasure in observing country life and wild nature. In winter it is hard to keep still in the cold. But the role of the silent watcher is even more important in winter than in summer. The rewards are often faster and richer. Thinking I saw a snowshoe hare behaving in a very curious and half-paralyzed fashion one day in December, I sat down in a swamp with my back against a thick cedar, and kept absolutely still.

In twelve minutes, I saw the following things: a fox passed thirty feet from me, completely unaware. A flight of over thirty red crossbills landed in a hemlock tree a few yards away, investigated the cones and flew off. A shrew, which weighs less than a ten-cent piece, came out from under some bark and leaves on the ground six feet from my boot, put on the most extraordinary shimmy dance, went backwards, went forward, went sideways, shook, shivered, twitched and vanished. A Canada jay, rare in these parts, softly flew in, with a low whistle, and had a long, amazed look at me from a distance of ten feet; and then, deciding I was alive after all, flew regretfully away.

Then something in the white snow caught my eye. It was the half-paralyzed snowshoe hare coming round again on the same track on which I had seen it before. It was more paralyzed, more laboured than ever. Its ears waved. Its eyes appeared half closed. It took three constricted hops, then hunched up, its ears laid straight back. Suddenly, it gave a huge leap. But the next hops it took were the same stiff laboured ones as before.

Out of the corner of my eye, I saw a small black object moving off to the right. It was the tip of the weasel's tail. Coming busily and steadily in the snowshoe hare's tracks, the weasel was taking long jumps and writhing runs through the snow. It was fated, of course, to win the chase.

All this in twelve minutes of motionless silence, eighteen miles from the core of a big city.

Muskox

In nature, sometimes the thing that is designed to be the greatest defence turns out to be the greatest weakness. The muskox is the most tragic example. This rugged wild ox, which has survived from time immemorial the most savage climate of the Arctic, was given as its basic instinct for defence the trick of backing together into a solid circle, whether the herd was five or fifty oxen. And with their massive heads, armed with sharp curved horns, forming a solid wall against any enemy, they stood off the mighty Arctic wolves and goodness knows what forgotten and long-extinct marauders of the past.

The Indians and Eskimos held the muskox in supreme regard, because their steaks are unrivalled, their skins and luxuriant hair, their masses of rich fat, their great heavy horns, were the materials of survival in that stern Arctic land. But it took a lot of courage and extraordinary equipment to go up against that wall of horned heads, from which, at the slightest opportunity, a bull as nimble as a Jersey would leap to attack anyone approaching near enough; and as nimbly would leap back to his place.

But when the Indians and Eskimos got modern rifles, a herd of muskox was completely at their mercy. They did not run, as other large game does. They stood fast, and were easily slaughtered, standing.

Eskimos were not the only destroyers of the muskox. Victor Cahalane, in his *Mammals of North America,* reckons that the various Peary expeditions to the Arctic accounted for over six hundred of these great animals, whose numbers are limited in the first place by the severity of their habitat.

The museums of Europe and America were offering thousands of dollars for a muskox, and to get them, the animal hunters had to kill a whole herd, which they did with the greatest ease, in order to capture the calves.

Lately, the muskox, sadly reduced in number and vanished from a great many of his former ranges, has been rigidly protected by the Canadian government. But this relief was almost too late, because of the muskox's conviction, supplied him by nature, that he was unassailable.

Owls

Few families in nature have a greater variety of size than the owls. There are owls no bigger than a baked potato, like the pygmy and the saw whet. But the great grey owl has a wing spread of over five feet.

Mid-winter is the time of year to see some of the most remarkable members of the family. Both the snowy owl and the great grey are visitors from the Arctic and sub-arctic during the winter, and they come down and find our January very comfortable, whether in the Maritimes or British Columbia, and everywhere in between. The beautiful snowy owl is likely to be found perched on a fence post, haystack, or even on a signboard within the city limits. The great grey sticks to the woods, like the other big owl most likely to be seen in bush, the great horned owl. The myth that owls cannot see in daylight is easily disproved by scaring one of them. And it flies through the tangle in daylight as easily as any crow.

The way to find owls is to poke about quietly in a cedar or spruce swamp, watching on the snow under the trees for the curious grey "pellets," about the size of cigar butts, which are the regurgitated skin, bones and other less digestible parts of the mice the owls eat whole or in lumberjack lumps.

Where you spot these "pellets," an owl is likely to be sitting a few feet above, close to the trunk, and looking extremely preoccupied.

One of the most attractive owls to find is the short-eared, about the size of a crow, of a particularly beautiful buff and ochre colour, striped and blotched; and its strange, staggering, moth-like flight is most engaging to behold. In summer, it haunts marshes. In winter, you might see it hunting in daylight around suburban fields where mice abound, or near golf links. Finding an owl in an evergreen tree in winter has some of the aesthetic shock of finding an orchid in June.

By the end of February, the great horned owl is nesting, and the search of a good big woodlot might be rewarded with the discovery of one of these unseasonable nests in an old crow or hawk nest. If you hear crows making a public protest out of all keeping with their own villainy, you will know they have got an owl or a fox in the crow pillory.

Owls have an age-old reputation for wisdom. But their effect on me is one of undiluted comedy. Maybe a story of an owl I heard in youth has affected my opinion. An Irishman who wanted to buy a parrot was sold an owl by an unscrupulous pet dealer. After he had the owl at home a few days, a friend enquired if the parrot was talking yet. "No," said the Irishman, "but he's doing a hell of a lot of thinking!"

Nothing better describes the absurd air of concentrated thought which characterizes all owls. The sorriest aspect of owls is their loneliness. Socially, they seem to be the outcasts of the bird world. They are invariably sitting alone, in a dark place, as though doing penance. Ducks, gulls, congregate in talkative companies. Even the hawks fly about in some companionship, and form great convoys for their migration. All the smaller birds move and live in communal association. But the owls seem to have gone apart to figure something out. It might be that they just don't like the flavour of mice to which provender they are condemned, and are simply sulking against nature.

Thought for February

Somewhere between twelve and fifteen billion wild birds – many of them song birds, all of them either beautiful or astonishing, are right now starting from Central America, Mexico, the West Indies and the southern United States for points north.

Some billions will stake out their claims to nesting sites all across the United States. Several billion, however, will keep right on coming north into Canada and on north up to and into the Arctic Circle.

It is a sort of a tidal wave of birds. Though it is still ice-bound here in Canada, the wave is already building up and sweeping across the far south edge of the continent. It will not engulf us until May.

Like a tidal wave, it pushes ahead of it smaller waves; the horned lark, for instance, from Halifax to Victoria, is on the wing, its more northern members heading for the polar islands, its southern members arriving to nest right here, often amidst the snowdrifts.

The estimate of between twelve and fifteen billion birds in North America is that of the National Audubon Society. For years, they have been counting them. The scientists say that if anything should prevent the birds coming north – say an atomic bomb or bacterial warfare attack on the birds in their distant congregations – our civilization would vanish in one year under an indescrible plague of insects, rodents and trash vegetation.

It may be so. The economic attitude is taking possession of our thoughts on every conceivable thing. But it is a horrible even a blasphemous way to look at birds. If the Creator put birds in the world solely for the economic security of greedy, anxious little man, why did He make them so beautiful and fill them with such incomparable music?

See how short the winter has been?

The real winter, I mean. How busily it hustled by. December all arush. Then January clanging by, with bobtail February hastening through. Then March, where we caught ourselves watching, out of the corners of our eyes and hearts, for the first wedges of Canada geese, heading where?

North, of course, where we'll be soon. There is an illusion amongst us Canadians that winter is a long, long business. We make jokes about the two seasons: July-August, and then winter.

It is preposterous. We have in Canada a longer summer than almost any place on earth. The long, hot summer! Now, I grant you that it is possible, with a calendar in front of us, to divide the Canadian year into clear-cut sections, each two months in duration. January-February, deep winter. March-April, late winter, spring on the threshold. May-June, spring and the jelling of summer. July-August, summer. September-October, autumn. November-December, the promise of winter.

But truth is not demonstrable by mathematics or logic. Truth is in the heart. And summer is that over-powering, over-mastering season, from May to late October, which goes on and on and on, as if it will never end, and dwarfs all other seasons, in memory, to mere interludes.

Greg Clark in Totems

Greg Clark's whimsical glimpses of life have provided some of the most enjoyable reading in Canada. Totem are delighted to announce that they are publishing a collection of six books by Gregory Clark — all for the first time in paperback. Watch for them!

THE BIRD OF PROMISE $1.95
Available now

OUTDOORS WITH GREGORY CLARK $1.95
Available now

GRANDMA PREFERRED STEAK $1.95
Available January 1978

FISHING WITH GREGORY CLARK
Available Spring 1978

THINGS THAT GO SQUEAK IN THE NIGHT
Available Spring 1978

MAY YOUR FIRST LOVE BE YOUR LAST
Available Fall 1978

For other Totems see overleaf